The Chimpanzees of Rubondo Island

I0131712

How did a random batch of chimpanzees come to populate a small island in Tanzania where apes had never lived before? Combining information gathered from fieldwork, laboratory and archival research, this book tells the unique story of chimpanzee babies taken from their forest homes in West-Central Africa and sold to European zoos and circuses, to then be shipped to Lake Victoria and set free on Rubondo Island. These founder animals learnt what to eat, how to build nests, to breed and raise young – ultimately forming a chimpanzee-typical fission-fusion society that today is thriving. The authors compare the ecology, behaviour and genetics of the Rubondo population with communities of wild chimpanzees, providing exciting insights into how our closest relatives adjust to changing environments. At the same time, a reconstruction of the historical context of the Rubondo experiment reflects on its chequered colonial heritage, and the introduction is viewed against current threats to the survival of apes in their natural habitats. The book will be of interest to scholars and professionals working in primatology, animal behaviour, conservation biology and postcolonial studies.

Josephine Nadezda Msindai obtained a BSc in Biological Sciences from King's College London, UK (2005), an MSc in Primate Conservation at Oxford Brookes University, UK (2008) and a PhD in Anthropology from University College London (UCL), UK (2018). *The Chimpanzees of Rubondo Island* is based on her doctoral work that included almost two years of field research in Tanzania.

Volker Sommer is Professor of Evolutionary Anthropology at University College London (UCL), UK. He obtained his PhD in Anthropology at Göttingen University, Germany (1985) and has conducted extensive primatological field studies in India (since 1981), Thailand (since 1984) and Nigeria (since 1999).

The Chimpanzees of Rubondo Island

Apes Set Free

Josephine Nadezda Msindai and
Volker Sommer

Routledge
Taylor & Francis Group

LONDON AND NEW YORK

First published 2023
by Routledge
4 Park Square, Milton Park, Abingdon, Oxon OX14 4RN

and by Routledge
605 Third Avenue, New York, NY 10158

Routledge is an imprint of the Taylor & Francis Group, an informa business

British Library Cataloguing-in-Publication Data
A catalogue record for this book is available from the British Library

ISBN: 9780367422196 (hbk)
ISBN: 9781032329178 (pbk)
ISBN: 9780367822781 (ebk)

DOI: 10.4324/9780367822781

Typeset in Sabon
by Deanta Global Publishing Services, Chennai, India

Contents

Illustrations

Figures

Box Figures

Tables

Acknowledgements

For inspiration to work on Rubondo Island, JNM is grateful to Klara Petrželková while VS is indebted to Gustl Anzenberger.

The authors appreciate all those who supported this book project, including commenting upon sections and sharing results of their own unpublished research: Gustl Anzenberger, Simon Bearder, Lucia Bobáková, Monica Borner, Frands Carlsen and Tom de Jongh (chimpanzee EEP coordinators), Helen Chatterjee, Katharina Herrmann (EAZA), Michael Huffman, Matthias Klapproth, Mwanahamisi Issa Mapua, Wolfgang Matschke, Vladimir Mazoch, Liza Moscovice, Guido Müller, Klara Petrželková, Felix Schürmann, David Spratt and Catryn Williams.

For invaluable help during fieldwork on Rubondo, JNM wishes to thank Markus Borner, chief park warden Massana G. Mwishawa, Fredrick Joakim Chuwa, Wickson Kibasa, Hobokela Mwamjengwa, Kevin Nkulila. Asante sana to: James Leonard, Joseph Mgwesa, Simon Nkuzi, Onesmo Mazazele, Samson Obiedy, Samuel Sudi, Aditi Pophale, Paco Bertolani, David Dietz, Musa Mwenge and Damali Obiedy.

Genetic analyses at the Deutsches Primatenzentrum, Göttingen, were kindly enabled by Christian Roos, Christiane Schwarz and Jinchuan Zhao. Images were contributed by Kat Brinsley, Sibylle Haase and Fritz Haase, Paul Kivuyo, William Rutta, Robert Slater, Zoo Basel (Switzerland) and Okapia Bildagentur (Christian Grzimek, Joachim Moog). Major financial support came from Asilia (African Safaris and Travel) via the Honeyguide Foundation, with special thanks due to Damian Bell, Allan Earnshaw and Jeroen Harderwijk. Short-term grants were provided by the North of England Zoological Society/Chester Zoo and the Faculty of Social and Historical Sciences, UCL.

Permits for fieldwork on Rubondo Island National Park were granted by the Tanzanian authorities (COSTECH, Commission for Science and Technology, permit numbers: 2013-297-NA-2007-02, 2013-297-NA-2007-01, 2012-186-NA-2001-18; TAWIRI, Tanzania Wildlife Research Institute; TANAPA, Tanzanian National Parks Authority).

JNM would like to thank her family, including her grandmother who inspired her interest in animal behaviour from a very young age; her parents for encouragement; and David Dietz for unflinching partnership. The Aiello lab group at UCL provided companionship and stimulating discussions, specifically Kathleen Bryson, Ally Palmer, Ella Al-Shamami, Matt Skinner and Christophe Soligo.

Postscript: While tracing the steps of the original Rubondo apes, JNM stumbled upon an astounding coincidence: 20 years after the chimpanzees had embarked on their journey from Europe to Africa, she had made that same journey on a cargo freighter from Europe via the Suez Canal to Dar es Salaam as a young girl roughly the same age as her chimpanzee namesake Joséphine.

Introduction

The authors of this book have worked in primate habitats from Morocco to Cambodia and Indonesia, Thailand, India, Nigeria, DR Congo, Uganda, Kenya and Cameroon, and have surveyed primates in Brazil, Argentina, Nepal and Malaysia. The one feature that connects all these places is the increasingly precarious status of non-human primates – as humans exterminate them from more and more of their natural habitats.

Chimpanzees are no exception. Across Africa virtually all populations – together with sympatric wildlife – are impacted by people. The list is long: small-scale logging; cattle herding; hunting for bushmeat, trade in pets or so-called traditional medicine; transmission of deadly diseases including anthrax or Ebola; habitat destruction to construct roads, dams, railways; conversion of forests for agriculture or palm-oil plantations; mining for minerals or precious stones; felling and burning of trees to produce charcoal (Arcus Report 2015, 2018). In short, people aren't kind to their closest relatives.

It is therefore ironic, that a 'human-made' population of apes – the Rubondo chimpanzees – should be able to live a life virtually unaffected by people. As primatologists familiar with the situation at other sites, we cannot help but feel a sense of relief while wandering across the island's forest. One doesn't stumble upon clearings where trees were recently felled; there are no scars in the hillsides, trodden by livestock; the smell of burning fires is absent; there is no rattling from motorbikes that plough the landscape; nor the searing noise of chainsaws.

This peace of nature has come at a price – most notably suffered by hundreds of human inhabitants who, in the 1960s, were ejected from their ancestral land. The remnants of a school erected a few months before their eviction still stands, as well as fruit trees like lemon, mango and avocado that happily continue to grow. Today, only park rangers and tourist lodge staff inhabit the island – roughly 120 people. These humans do not pose a threat to wildlife, they are there to protect it. And they do – aided by the fact that Rubondo is an island, fairly distant from the mainland.

How did a small cohort of chimpanzees come to populate an isle where apes had never lived before? To tell the unique story of the Rubondo chimpanzees, we combine results from research conducted in the field, in the lab and in archives. Kidnapped as babies by animal traders in African jungles and sold to Europe to be exhibited in circuses and zoos, the ancestors of the Rubondo apes faced a lifelong existence in captivity. However, in the early 1960s, the then director of the Zoo Frankfurt, Bernhard Grzimek – famous already for his Oscar-winning documentary *Serengeti Shall Not Die* – developed a plan to create a sanctuary for threatened animals on an island in Lake Victoria. Grzimek combined this idea with the conviction that wildlife conservation will only be successful if it can pay for itself, chiefly through tourism. Because visitors are attracted to places where big animals roam, Grzimek continued earlier initiatives of the outgoing British colonial wildlife authorities to translocate animals from Tanzania's mainland to the island reserve – which led to the introduction of iconic mammals such as rhinoceroses, elephants and giraffes.

In addition, back in Europe, Grzimek assembled 17 captive-held chimpanzees destined to be set free in the island's forests. Several of the ex-captives died, one en route; others were shot dead when attacking rangers and an unknown number likely perished in the unfamiliar new surroundings. Still, several survived, becoming founders of a future population. Left completely to their own wits, these apes, against all odds, learned to eat wild foods and how to construct night nests in trees, to ultimately breed and raise young and form a chimpanzee-typical fission–fusion society.

A comparison of the ecology, behaviour and genetics of the Rubondo apes with endemic wild chimpanzees provides exciting insights into how our closest relatives adjust to diverse environments.

Nevertheless, in hindsight, questions need to be asked about the modus operandi of the Rubondo introductions. How do these chimpanzees, with no preparation or post-release support, having been literally 'dumped' into wild and unfamiliar surroundings, compare with other ape translocations? How does Grzimek's unorthodox experiment hold up against contemporary considerations of animal welfare and animal rights? Moreover, applying modern guidelines of animal introductions, should we consider the Rubondo releases a failure or a success? And finally: How is the Rubondo experiment embedded in a chequered heritage of colonialism?

No matter what – while elsewhere, competition with humans increasingly tests the limits to ape adaptability and drives our evolutionary kin ever closer to extinction, this artificial colony on Rubondo island has become a chimpanzee haven.

1 Creating a Wilderness: The Making of an Island National Park

Figure 1.1 View of Rubondo Island from a hilltop. (Photo: Rob Slater, Safari Consultants)

The shores of the mighty Lake Victoria and its islands were once entirely covered with trees. Today, this ancient vegetation has all but disappeared. A conspicuous exception, tucked away in the lake's south-western corner and still blanketed in lush green forest, is the island of Rubondo (Figure 1.1, Figure 1.2).

A natural paradise untouched by humankind is how visitors perceive Rubondo's evergreen rainforest. Tourism companies promote this national park as 'wild and virtually uninhabited'. Yet the island and its surrounding islets are by no means timeless, and this place does not lack tangible history.

DOI: 10.4324/9780367822781-1

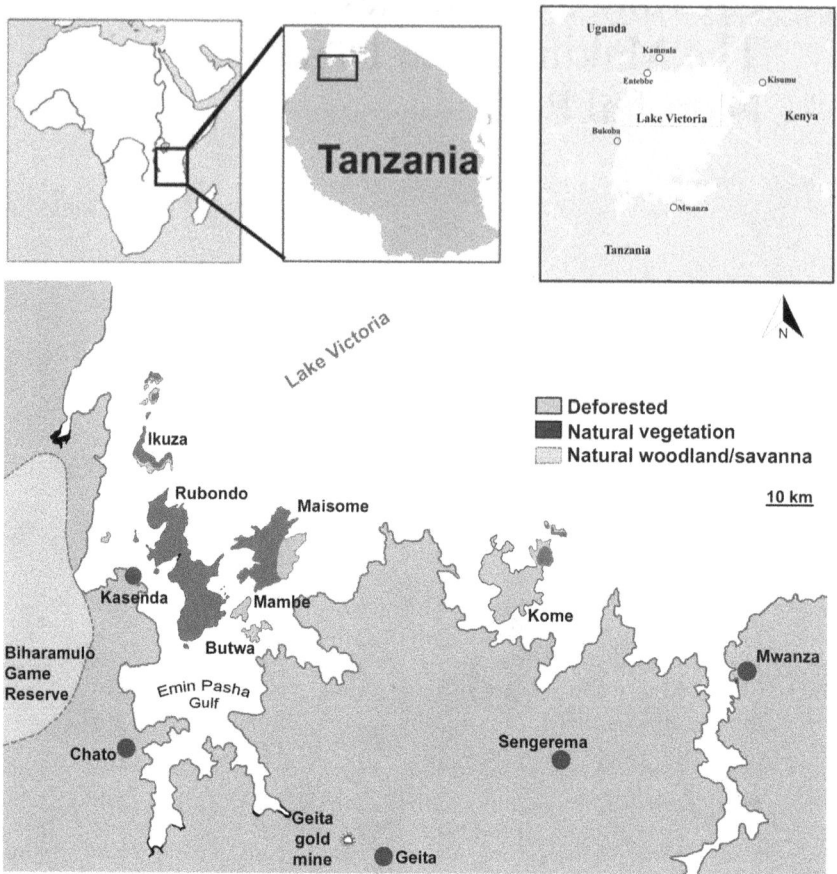

Figure 1.2 South-western Lake Victoria region, Tanzania, East Africa. (Map: JNM)

In fact, the supposedly pristine wilderness has been shaped over centuries by anthropogenic forces, particularly through recent interventions which took place in the 1960s and 1970s. These involved the introduction of large animals to the island, the resettlement of local people to the mainland and the construction of infrastructure for tourism.

Within this context, our book focusses on what we know about the historical – or shall we say historic – introduction of 16 chimpanzees that took place more than half a century ago. Here we trace the origins of the founder apes; how they and their descendants managed to survive and thrive in an unfamiliar ecology; and we'll reflect on what this experiment teaches us about the future of our closest living relatives in a now solidly established Anthropocene era.

A Jewel of Nature: Shaped by People

One person is central to our story: the charismatic and at times controversial German zoologist, nature conservationist, author and filmmaker Bernhard Grzimek (1909–1987). To situate the Rubondo releases, we chronicle his life and the socio-political landscape in which Grzimek operated in, because the story of how a bunch of chimpanzees came to live on an African island is imbued in a history of colonialism, decolonisation and neo-colonialism.

Bernhard Grzimek likened Rubondo to a "gem" (Grzimek 1971: 17). However, the German conservationist did not really think his metaphor through, because a gem is a stone made more precious through polishing – while Grzimek was a fan of nature uninterrupted by human actions. Still, anthropogenic involvement has left undeniable marks on Rubondo, and whether or not this ennobled the island is well worth debating. In any case, the chimpanzees of Rubondo are a shiny facet of this 'jewel' in Lake Victoria, a prominent part of the re-imagining of a once human-populated island as an untouched natural space (Schürmann 2017b: 285).

To render subsequent reading of chronological detail easier, here is the epitome. During the colonial German occupation of East Africa (ca. 1880–1918), the island's human inhabitants were forced to vacate Rubondo, allowed to resettle when the British took control of the region after World War I and ultimately evicted in 1964 (Kiwango et al. 2005). Already a forest reserve, Rubondo Island was gazetted a game reserve in 1965, and in 1977 became a national park together with 11 small adjacent islets and portions of the surrounding lake. Much of this development was due to lobbying by Bernhard Grzimek, director of the Frankfurt Zoological Garden in Germany from 1945–74 and president of the Frankfurt Zoological Society from 1971–87. Taking inspiration from British colonial wildlife officers, Grzimek sought to convert the island into a sanctuary for threatened animals, and he persuaded the newly independent Tanzanian government that tourism could be used to generate revenue for the management of such reserves (cf. Schürmann 2017b: 285). In conjunction with these ideas, various types of mammals were set free from 1963–73 to populate the island. The portfolio comprised rhinoceroses, giraffes, elephants, guereza monkeys, roan and suni antelopes. In addition, Grzimek sourced chimpanzees held in European institutions, which were released in four waves between 1966 and 1969. Finding themselves on a large, forested island, free from predators and hunting pressure, habitat destruction and conspecific competition, the apes, without further human support, formed an island population that now numbers in the dozens.

The general sketch of historical developments will be expanded in the following, except for the chimpanzee releases themselves which will be detailed in the subsequent separate chapters.

Scattered Records: Piecing Together a Story Untold

Today, Rubondo embodies African wilderness. However, this is only because of structural measures that came to pass in the 1960s and 1970s. These involved translocations and introductions of large animals, as well as the eviction of local people, processes that culminated in the creation of a national park in 1977.

The saga of the construction of a 'natural space' on Rubondo has so far only been told in fragmentary ways – and sometimes, as we'll discuss, incorrectly. An articulated account in English is entirely lacking. This is at least partly because its dynamics date back to the late 19th and early 20th centuries, when this chunk of East Africa was exploited by the German Empire, while the ecological refurbishment on Rubondo coincided with the transition between the end of colonialism and the first decade post-independence. These trajectories still reverberate when, in the decades after World War II, the Frankfurt Zoological Society (FZS) and its president Bernhard Grzimek became engaged here. Many publications, reports and archival sources are therefore written in German, constituting a barrier for those who do not speak the language. This also pertains to the most recent summaries of the events in the FZS's *Gorilla* magazine, which include some hitherto unpublished images of the historic events (Dinter 2021, Msindai 2021).

Bernhard Grzimek, as main protagonist, captured the journey of the first batch of chimpanzees from Europe to Rubondo in a 47-minute episode which aired on his TV show *Ein Platz für Tiere* on 15Nov66 (Grzimek 1966a). The film's narrative largely concurs with often similar printed versions about the release, of which only some were translated into English (Grzimek 1966b, 1966c, 1967, 1969, 1970, 1971, 1988). These accounts differ in their details, perhaps caused by tacit motives to present the events in the most favourable light.

A first systematic effort to tap into original records and correspondence about the Rubondo releases which had accumulated in the FZS archives in Frankfurt am Main was made back in 1973 when, in preparation for fieldwork at the island, the German animal behaviourist Gustl Anzenberger, then at the Institute of Zoology at Ludwig Maximilian University of Munich and the Max Planck Institute for Behavioral Physiology, Seewiesen, collated oral and written information. During his subsequent research visit to Rubondo in 1974, Anzenberger did not focus on chimpanzees, as they were too difficult to locate. Instead, he conducted research on carpenter bees (Chapter 2). However, once he moved to the Institute of Anthropology at the University of Zurich, Switzerland, Anzenberger's archival compilation informed a report about a 1994 field study on the island for which he had enlisted the young Swiss biology graduate Guido Müller.

Some 40 years later, the German historian Felix Schürmann, through archival visits in 2014, compiled his own independent data base of the events. As a member of the human–animal studies group at the University of Kassel, the Leibniz Institute for European History in Mainz and the Gotha Research Centre at the University of Erfurt, he explores the history of politics of wildlife during decolonisation. For his research on the history of Rubondo, Schürmann (like Anzenberger) consulted repositories at the Zoologische Gesellschaft Frankfurt (Afrika, 00001976 Rubondo 1, shelf 01019; Afrika, 00001970 Rubondo 3, shelf 01021), but also the Frankfurter Institut für Stadtgeschichte (Zoologischer Garten 195–199, 209–210, 212) and the Tanzania National Archives, Dar es Salaam (967.822.1 West Lake Province Annual Report 1960; Acc. 599, GD/14/4 Tanganyika Game Ltd.; Acc. 599, GD/14/18 Game Catching for Translocation and Re-stocking by Game Dept.; Acc. 599, GD/18/3 Tanzania National Park General; Acc. 599, GD/22/3 Senior Game Warden's Conference; Acc. 599, GD/25/3 Chief Game Warden's Safaris).

In our reconstruction of the events surrounding the release of chimpanzees onto Rubondo, we draw heavily on the reports (Müller 1995; accessible online as Müller & Anzenberger 1995) and publications (Schürmann 2016, 2017a, 2017b) informed by those archival explorations. In these sources, interested readers can find references to letter exchanges and documents, which we do not quote in detail. Importantly, we communicated directly with Gustl Anzenberger and Felix Schürmann, while comparing their independently compiled lists about the life histories of the chimpanzees shipped to Rubondo, and cross-checked some conflicting entries. All German-language sources were collated and translated by co-author VS.

Banyarubondo: The Original Islanders

Nowadays, the prominent inhabitants of Rubondo are its animals. However, far from being the unpeopled wilderness as the national park is now marketed, humans had lived on Rubondo for a considerable period. They went by the collective KiSwahili name of *Banyarubondo* – which simply translates into 'The People of Rubondo'. Unsurprisingly, their history and fate are entwined with the geographical and historical context of that part of East Africa they called home (Pakenham 1992).

From the late 19th century, colonialists associated with the German Empire (the 'Second Reich', created 1871) annexed regions that are now Tanzania (minus Zanzibar) and incorporated them into German East Africa (GEA). Once Germany was defeated in World War I, the Treaty of Versailles in early 1920 transferred large parts of GEA to Britain, with 'Tanganyika' becoming the name of the British territory. Tanganyika became independent and joined the British Commonwealth on 09Dec61, albeit it was overseen for one more year by a British governor-general. Shortly after, on 26Apr64, the Zanzibar archipelago merged with Tanganyika. Alluding to its two former parts, the new country was renamed the United Republic of Tanzania. Its populace grew from 10 million during independence to 60 million in 2020.

The Banyarubondo were subjects of these large-scale geopolitical upheavals. Their history, traditional lifestyle and use of natural resources were reconstructed by Tanzania National Parks' researcher Yustina Andrew Kiwango and colleagues, who, from 2003 identified and contacted former inhabitants of the island. They interviewed the Mzees ('old men') and knowledgeable 'experts', although their recollection of dates remains sketchy because "no one knew how to read or write so they do not know the exact years" (Kiwango et al. 2005: 5).

Nonetheless, as far back as the 1800s, Rubondo had been populated by Zinza-speaking people. At the time, the western shores of Lake Victoria were ruled by kingdoms including clans of the Karagwe and Businza (perhaps akin to Zinza). The Zinza-ruler Mukama resided in what is now the Geita region. Upon the death of his son Lukakaza, the Wahinda clan took control of Rubondo. During wars and conflicts, for example with the Baganda people, the Banyarubondo hid in caves such as those at Aharusumbe, as testified by finds of human remains (Kiwango et al. 2005: 8).

The German publicist Carl Peters (1856–1918) was the driving force behind initiatives to take colonial control of East Africa. In 1884, Peters founded the Gesellschaft für Deutsche Kolonisation (Society for German Colonisation; Speitkamp 2005), a pressure group for the acquisition of colonies. During the autumn of 1884 he visited East Africa, sought out local chiefs and – often after offering them copious amounts of alcohol – handed them German-language documents, on which the chiefs drew crosses as signatures. These documents promised protection from enemies and, conversely, granted the colonisation society unrestricted rights to levy duties and taxes, establish a judiciary and administration, bring in armed troops and give settlers the "mountains, rivers, lakes, and forests" to use as they wished (Schmidt 1898, Speitkamp 2005). These so-called 'Schutzverträge' ('protection treaties') were used to form the Deutsch-Ostafrikanische Gesellschaft in 1885. This smokescreen enabled the ultimate establishment of German rule, which lasted until the end of World War I, when control of the colonies of Deutsch-Ostafrika fell to the British.

Towards the end of the German occupation, in 1911, the Banyarubondo were asked to vacate the island (Schürmann 2021). The removal of people was in conjunction with efforts to recruit labour to eradicate the tsetse fly, a parasite that hampered the breeding and exploitation of livestock. At the same time, this forced displacement permitted Rubondo to be designated as a forest reserve (Tanzania National Parks 2003; cf. also Kreye 2021). Once the British took control of the region, the Banyarubondo were allowed to return. Their round houses had walls from mud and timber and grass-thatched roofs. Distributed widely around the island, they cultivated flat elevations along the coast, where, as the interviewees reminisce, they were blessed with fertile soil. Steep slopes and hilltops were left uninhabited, which kept deforestation to a minimum and prevented soil erosion (Kiwango et al. 2005).

The various subsistence activities enabled a vibrant trade between neighbouring islands and villages on the mainland. To prepare the land, fires were burnt at times, which also ensured fresh grazing pasture for domestic animals. People maintained small plots of banana, pineapple, maize, cassava, sweet potatoes and finger millet. Some of the surrounding small islets were also cultivated, including Izilambuba, which served as a banana plantation. Fruit trees provided orange, papaya, mango, lemon and avocado. The latter three types of trees can still be found fruiting on the island to this day. "A white man", sometime in 1937, "brought with him Senna seedlings" – and thus introduced *Senna spectabilis* to the island. The leguminous tree, which blooms conspicuous golden-yellow flowers, is a native to South and Central America. Likewise known as American cassia, Golden wonder tree, or Popcorn tree, the quickly growing plant is treasured for its ornamental properties while also serving as a sunshade as well as a good source of firewood and local building material. The current management aims to eradicate alien plant species inside the national park, especially very invasive taxa such as *Senna spectabilis* (ibid., p. 9).

Apart from farming, the Banyarubondo engaged in fishing, hunting and gathering. Just south of Kageya, the 'Holy Waters' wetlands were very clear and calm, and functioned as a wetland fishery, with harvest tied to fluctuations in water levels. A favourite game was the sitatunga antelope, which roamed these swampy areas thanks to their splayed hoofs. From the 1960s, the lake level dropped by around 2 metres (Mnaya & Wolanski 2002, Mwanuzi 2006) and this once wetland has all but dried up today. Exploiting the hives of wild bees was also commonly practised. The forest allowed the gathering of useful plants such as grains of paradise (*Aframomum* sp.), the rubber vine known locally as amakamila (*Saba comorensis*), sharazi (*Garcinia huillensis*), amatemerere (*Pancovia turbinata*), ensuungwa (*Vitex mombassae*), and the edible roots of amasoome and enchulile – florae, most of which we will re-encounter as important components of chimpanzee diet (Chapter 5).

Education was informal. The adult men raised boys to display "good habits, industriousness and respect", and trained them how to fish and farm. Mothers and grandmothers taught girls how to cook and prepare traditional medicines. From age 14, young men versed in traditional medicine could inherit their fathers' title of medicine man. A female healer, the Emukabela, officiated as priestess at communal rites (Kiwango et al. 2005: 5, 16).

In the 1930s and 1940s, Christian missionaries of Catholic and Lutheran conviction arrived. While schools were already constructed in the wider area, it was not until 1964 that one was erected on Rubondo at Lukaya. However, schooling was short-lived, because about three months later, "a command from the government required all people living in Rubondo and surrounding islands to vacate the area, and never to return" (ibid., p. 4) – a development related to the envisioned conversion of the island into a nature reserve.

Today the only tangible memory of the Banyarubondo is a kaput school building. But, while the original islanders are gone, the names they gave to particular places are still in use. There is 'shuleni', an echo of the German word 'Schule' for a site where the first colonial rulers had built a primary school. 'Lukaya' translates as 'the chief's seat', and 'Mlaga' as 'to bid a person farewell', given that it was from here that the chief went on trips to his territories on the mainland. Other names refer to natural properties, such as 'Kasenye', which means 'sandy area', or 'Masekela' for many 'misekela' plants or 'Lukukuru' for its richness of 'enkukuru' trees (ibid., p. 44).

When the Banyarubondo were directed to resettle, many preferred to go to villages where they had a familial connection, including on Ikuza and Maisome islands as well as Izumacheli, Nyabuzera, Kisaba, Butwa and Nkome on the mainland. Some people continued to live on Rubondo, working for the new park management in, for example, road construction. The last Banyarubondo left in the early 1980s (Kiwango et al. 2005). Their formerly cultivated plots "have slowly reverted to a wild state" (Rodgers et al. 1977).

Each island family maintained its own sacred site where they believed their ancestor's spirits lived. The Banyarubondo clans tried to transfer them to their new homes, but the spirits would not leave and, despite traditional rituals, somewhere on the way the spirits returned to their old abodes. Elders were sometime allowed to come back to the island to conduct ceremonies and sacrifices. During some of these rites, families said goodbye forever to their spirits: "We are sadly leaving you, we will not be coming back because new people have come into this place and we will not be allowed to visit you again" (Kiwango et al. 2005: 66).

Yet, the origin of the name 'Rubondo' itself remains a mystery. One theory maintains that it is a developed pronunciation of the Zinza-word 'Rwamalebhe', which signifies wealth and power. However, 'Rwamalebhe' is also the name for a small pond in the island's south. It refers to 'amalebhe', the fine, always present vegetation floating on the surface of the water. The natives revered this place as they believed it to be the centre of the island, far from anywhere people dwelled, about 7 kilometres from the shore. It was believed that evil men or those who set forth purposely to search for it could not see it. The pond is well hidden in the forested Nyakutukula hills. That name, derived from 'Nzendabahakhile', incidentally makes for a good motto for any researcher tracking the elusive chimpanzees, given that it means "eat to your fullest first because this forest is vast" (ibid., p. 23).

Grzimek: Famously Hard to Spell, but Rightly Famous

Who was that man with a family epithet difficult to expound and enunciate – even for German speakers? Why did he become a household name in Germany synonymous with animals and nature conservation – akin to the 'national treasure' that is David Attenborough in Britain? And why did

Grzimek intervene in the wildlife management of a newly independent East African nation?

Bernhard Grzimek was born in 1909, the youngest of six children, in Neisse in Upper Silesia, what was then part of the Second Reich of Germany (for the following, see Sewig 2009). Grzimek studied veterinary medicine in Leipzig and Berlin from 1928–33, submitting a doctoral thesis on the artery system of domestic chickens. When Hitler rose to power in 1933, Grzimek joined the Nazi militia SA and later also the NSDAP party. Throughout the Third Reich, Grzimek worked for government agencies, primarily concerned with disease control in cattle and poultry, as well as, improving the storage of chicken eggs.

In 1945, at the end of World War II, Grzimek moved to Frankfurt, a major city in Central Germany on the banks of the Main River. Here, because of personal connections, he secured the position of director of the bombed-out zoological garden. In the years and decades to follow, he rebuilt and reshaped the institution. Grzimek initially travelled to Africa in the early 1950s to acquire animals for his zoo. The first trips were in 1951 to French West Africa (today's Ivory Coast), to source chimpanzees, and in 1954 to Belgian Congo (today's DR Congo), to export the elusive Okapi. During his excursions, Grzimek became increasingly aware of the threats wild animals faced due to habitat destruction and big-game hunting, a realisation which spurred his efforts to protect nature.

Grzimek described his passions and concerns in popular books such as *Flug ins Schimpansenland* (1952; translated as *Doctor Jimek, I Presume*, 1955) and *Kein Platz für wilde Tiere* (1954; *No Room for Wild Animals*). Grzimek also skilfully employed the emerging new mass medium of television. In Germany, he holds legendary status because of his TV series *Ein Platz für Tiere* ('A Place for Animals'). The show aired 175 times between 1956 and 1987, often on a prime-time slot. An avuncular Grzimek knew how to captivate his audience, not least by featuring animals from his zoo, whether a young cheetah, serval, spider monkey, sloth, gorilla or orangutan. At the end of each episode, he asked viewers to donate money to an account entitled *Hilfe für die bedrohte Tierwelt* ('Aid for Endangered Wildlife'). A 2009 stamp commemorates the TV series (Figure 1.3). The show inspired many young people to pursue their own careers as zoologists – a fate that, incidentally, also befell the co-author of this book.

While Grzimek engrossed his audience with stories and films about animals, he became an increasingly outspoken conservationist. For this, together with his son Michael, Grzimek ventured into cinematography. To enable footage from the air, they both obtained a pilot licence and bought a single-engine light aircraft, painted with what became a trademark zebra stripes pattern. Based on the 1954 Congo expedition, the father–son duo first produced *No Room for Wild Animals*, a 1956 colour documentary film, that, unexpectedly, earned them awards and a windfall of money when sold to 63 countries. The Grzimeks offered to buy up parts of the Serengeti with

Figure 1.3 Bernhard Grzimek's iconic TV series *Ein Platz für* Tiere commemorated
on a stamp of the Deutsche Bundespost released for his 100th birthday
in 2009. (With permission from designers Sibylle Haase and Fritz Haase;
Bundesministerium der Finanzen, Germany)

the proceeds from the book and film. This move was meant to counter plans
of the last British governor-general of Tanganyika to cut off a large slice
of the Serengeti National Park to provide the pastoralists Masai with an
opportunity to graze their cattle and compensate for it by annexing other
areas. The British administration refused the Grzimeks' offer. However, the
director of the Tanganyikan national parks, Peter Molloy, suggested spending the money on a new survey of the migration routes of wild animals to
better establish the borders of the Serengeti. This led to extensive aerial surveys of herbivore herds and their migration routes through the vast plains
to the west of the Great Rift Valley. The Grzimeks realised that the annual
migrations went through parts that were planned to be degazetted, while the
proposed substitute area was hardly used. However, the colonial administration did not wait for the Grzimeks' results and decided that the northern
plains should be added to the park as planned. The eastern part was declared
to be the Ngorongoro Conservation Area, which the Masai were offered
as compensation and where they would be allowed to graze their cattle.
The conflicts surrounding these trade-offs became the subject of the 1959
documentary *Serengeti darf nicht sterben* (*Serengeti Shall Not Die*). It was
the first German film after World War II to win an Oscar, i.e. the Academy
Award for Best Documentary Feature, in 1960. The film's mission, to secure
a future for an ecosystem that harbours iconic herds of wildebeest, zebra and
gazelles plus the predators that pursue them, is still mired in disagreements
over land-use rights, particularly involving the pastoralist Masai (Grzimek &
Grzimek 1959, Homewood et al. 2009). Still, the film elevated the Serengeti
to celebrated status, a global prominence that has not waned since.

Increasingly, Grzimek not only became a conservationist, but also a cham-
pion of animal welfare, causing outrage with the German public when screen-
ing gruesome footage of the clubbing-to-death of baby seals to skin off their
prized fur, as well as protesting against the industrial production of chicken

eggs which confined hens to tiny cages. Grzimek also became instrumental in the establishment of the first German national park in the Bayerischer Wald in 1970. From 1970–73, Grzimek acted as the West German government's commissioner for nature conservation. In 1975, together with 20 like-minded environmentalists, he founded the non-governmental organisation Bund für Umwelt und Naturschutz Deutschland (BUND; Friends of the Earth Germany). In addition, from 1967–74, Grzimek was editor-in-chief of the 13-volume encyclopaedia *Grzimeks Tierleben* (Grzimek 1967–72). The monumental work saw translations into several languages, including the English version *Grzimek's Animal Life Encyclopedia* (e.g. Grzimek et al. 2003).

Grzimek, apart from being the director of Zoo Frankfurt from 1945–74, also presided over the Frankfurt Zoological Society from 1971 until his death in 1987. Founded in 1858, the Zoologische Gesellschaft Frankfurt originally lobbied for the creation of a zoo, which materialised in 1874. While the city of Frankfurt became the official owner from 1915 onward, the society was reinvented in the 1950s as an association for supporters of the zoo. Within its framework, Grzimek created the already mentioned special account to aid nature conservation projects. In 2001, the accumulated wealth of the FZS, some 33 million euros, constituted the start-up capital of the foundation Hilfe für die bedrohte Tierwelt, which aims to preserve wild places and ecosystems across the globe, but operates specifically in East Africa (fzs.org/en/).

Fuelled by his media appearances, Grzimek's turbulent and at times tragic private life has also garnered considerable attention. Married in 1930, he had three sons with his first wife Hildegard Grzimek (née Prüfer, 1911–84): Rochus (*1931), Michael (1934–59) and adopted son Thomas (1950–80). In Jan59, a plane piloted by Michael during the filming of *Serengeti Shall Not Die* collided with a vulture. Michael lost control and was killed in the crash. He was buried the same day on the rim of the Ngorongoro crater. The government of Tanzania later erected a stone pyramid over his grave. Thomas Grzimek, increasingly depressive and drug-addicted, took his own life in 1980. Grzimek's first marriage ended in divorce in 1973, and in 1978, he married the widow of his son Michael and his daughter-in-law, Erika (née Schoof, 1932–2020). He adopted his son's children Stephan Michael (*1956), and Christian Bernhard (*1959, after Michael's death). Christian later became the head of the 'Okapia KG Michael Grzimek & Co.', a photo agency specialising in wildlife imagery founded in 1954 by his father and grandfather. Bernhard Grzimek had two more children, Monika Karpel (*1940) and Cornelius Grzimek (*1945), from a long-term extramarital relationship, and there are rumours of love-children in Africa. Bernhard Grzimek passed away in Frankfurt am Main in 1987, when, while watching a tiger act at Circus Althoff, he fell asleep and suffered a heart attack. The ashes of the 78-year-old were buried next to his son Michael at Ngorongoro. After Grzimek's death, inheritance disputes made headlines for years. In 2015, Germany's channel 1 TV aired a three-hour long biopic. The name Grzimek, however difficult to pronounce and spell, is widely recognised in Germany till date.

Until the 1970s, nature conservation and environmental protection, in German-speaking countries and elsewhere, had remained hobbies of a bunch of idealists. It certainly is to Grzimek's credit, that 'green thinking' exploded soon after – up to the point that, by the 2020's, the political green party has a realistic chance of nominating Germany's chancellor. Twenty years after his death, a leading German newspaper summarised Grzimek's influence with the simple words: "Durch ihn wurden wir alle grün" ('Because of him, we all became greens') (Miersch 2007).

While Bernhard Grzimek's professional achievements are generally lionized, particularly by the institution he shaped so much, Frankfurt Zoological Garden, there are also critical voices regarding his actions and motives (cf. Sewig 2009, Schürmann 2017a). For one, after the end of World War II, while his membership in Nazi organisations, rather than driven by ideological conviction, had most likely been career opportunism, Grzimek persistently lied and systematically misled the authorities about this chapter of his life. Proof of his membership was only discovered in 2009, decades after his death – a stain on his career that caused the media which Grzimek had so skilfully instrumentalized to become less hagiographic (e.g. Kellerhoff 2015).

Also, while highlighting the damage colonial hunters inflicted on Africa's wildlife, Grzimek exploited those ecosystems as well when he acquired animals for display in zoos. Just for scale: Kidnapping one baby chimpanzee from the wild means accepting the death of roughly ten other conspecifics massacred during the hunt for the youngster (Hughes et al. 2011; see Chapter 6). Moreover, while Grzimek campaigned against certain types of animal cruelty, such as the battering of seal pups to strip them of their fur, and the misery of caged chickens, he condoned the exploitation of wild animals such as elephants, tigers or orangutans in circuses and in zoos – practices that involve considerable pain and suffering (Goldner 2014, Diehl & Tuider 2019). The idea of individual 'animal rights' remained foreign to Grzimek. This is also evident in the way he handled the deaths of some of the chimpanzees destined for release on Rubondo Island – facets we will revisit later.

In a larger context, there is increasing scrutiny about how Grzimek's actions are embedded in the dynamics of colonisation and decolonisation (cf. Schürmann 2016). Grzimek romanticised Africa to his largely Western and in particular German-speaking audience as a home of pristine naturalness. This alleged animal paradise, brimming with charismatic large mammals, is portrayed as beleaguered by trophy hunters and the destructive consequences of catch-up modernization along Western lines.

In continuation of a colonial trope, Grzimek labelled supposedly unspoilt parts of Africa as an "ideal possession" for all humanity. The title of a book embodies this view: *Rhinos Belong to Everybody* (Grzimek 1962). This attitude is similarly reflected in the famed concluding narration of the 1959 *Serengeti Shall Not Die* documentary:

> These last remnants of African animal life are the common cultural property of all mankind, just like our cathedrals, like the ancient buildings

[...]. A few centuries ago, Roman temples were demolished to build houses from the ashlars. Today, if a government [...] dared to demolish the Acropolis in Athens to build apartments, an outcry of indignation would go through all civilized humanity. Similarly, neither black nor white people should touch these last living cultural treasures of Africa.

The Serengeti plains became the focus of Grzimek's efforts to restrict access of local people, above all the Masai. As he narrowed the problem to a spatial conflict, his demands have been viewed (Schürmann 2017a) as echoing sentiments of pre-World War I German expansionism in Africa and the later Nazi ideology of 'Lebensraum' ('living space') (see also Gissibl 2019). This seems like an extreme take on the global visions of a wildlife enthusiast. However, Grzimek's concerns are certainly rooted in German history. As a result of World War I, the nation had its borders truncated and its growing population was perceived as a 'Volk ohne Raum' ('People without Space'). Grzimek simply turned this phrase on its head, by envisioning instead a 'Raum ohne Volk' ('Space without People'), i.e. "a place where animals could find refuge from burgeoning human populations who were using up more than their share of the Earth's resources" (Lekan 2016: 75). Drawing attention to the growing number of humans, he stamped letters with the Latin phrase *ceterum censeo progeniem hominum esse diminuendam* ('Furthermore, I am of the opinion that human progeny must be reduced'). Given that Grzimek perceives humans "as rogue mammals caged by self-created walls of modernity" (ibid., p. 77) and the rapid speed with which Africa became decolonised and modernised, he was afraid that the continent might "recapitulate Europe's pathological move toward modern ways of life — urbanized, brutalized, alienated" (ibid., p. 79).

Grzimek's Mathusian ecology, in some ways, continued to perpetuate the "myth of wild Africa" which had "authorized a green-imperialist 'scramble for Eden'" (ibid., p. 61). Indeed, where he speaks of people, a vocabulary of colonial clichés abounds. Grzimek's Africa is populated by dancing "tribes" and trusting "jungle children". Here and there one must beware of "cannibals", but a "faithful black boy" is always at hand (cit. in Schürmann 2017a). Many a bare-breasted village beauty is mentioned, and the countryside praised because "jet-black girls [...] went about gracefully erect with their round breasts serenely exposed" (Grzimek 1971: 203). While such utterings are nowadays considered racist, Grzimek also held outspoken progressive views. Thus, he goes to great length to engage with and debunk claims (whether made by the Nazis, Stalinist Soviets, European colonialists or Allied Occupation forces) that certain groups are less intelligent or cultured (whether Russians, Jews, Germans or "black Africans") (Grzimek 1974: 385–388, 460–461).

During Grzimek's forays into Africa, the pace of decolonisation picked up, with eight independent states emerging from 1960–64 alone. Grzimek's by now established fame gained him access to African heads of state and government. In what could be called strategic opportunism, he avoids statements that could place him in the general political spectrum. He often meets with Tanzania's left-wing President Julius Nyerere (1922–99) (Figure 1.4), but later as well with Uganda's right-wing dictator Idi Amin (1925–2003), becoming appointed to the board of trustees of the national park administrations of Tanzania and Uganda (Sewig 2009).

Grzimek emphatically appeals for trust in the black elites – especially in Nyerere who adopts his view that nature can be made to pay for its survival by attracting tourists, generating resources that will benefit the masses. And Grzimek scolds his own country: "Young African nations are shaming the older countries of the white man, in which many species have become permanently extinct." He finds this "especially humiliating to a citizen of the German Federal Republic" where politicians have yet to establish a single national park. Whereas, "in ten or twenty years' time, forward-looking statesmen like Dr Nyerere will be acclaimed as pioneers of a new cultural approach" (Grzimek 1971: 10).

Grzimek's network on the ground are white bureaucrats of British origin who initially remain in their posts after independence. He develops rapport with Myles Turner, chief game warden for the Serengeti, Bruce Kinloch, director of the Tanzania Wildlife Authority, and John Owen, director of the Tanzania National Parks Authority. Together with Kinloch, Grzimek in 1963 established the College of African Wildlife Management on the southern slopes of Mount Kilimanjaro above the city of Moshi. In 1961, Grzimek had already founded the Serengeti Research Project – which would go on to

Figure 1.4 Tanzanian President Nyerere (r.) and Bernhard Grzimek (l.) in 1963. (Photo: Okapia)

expand, along with various name changes (1966: Serengeti Research Institute; 1980: Serengeti Wildlife Research Institute; 1999: Tanzania Wildlife Research Institute). Both institutions contributed significantly to the institutionalization of Western concepts of nature protection and till date train Tanzanians and conservationists from other African nations in wildlife management.

The 1960s were Grzimek's heyday in East Africa, the decade also when he conceived and executed the Rubondo project – as we will elaborate upon in the next chapter. Grzimek scaled back his presence on the African continent in the 1970s, because of increasing new obligations for environmental causes in the Federal Republic of Germany. Moreover, his network in Africa was crumbling. Progressive 'Africanization' saw the end of post-colonial conservation bureaucracy, with Turner, Kinloch, Owen and other warrant officers losing their posts. Given cultural and generational differences, Grzimek could not find a common level with their successors (Schürmann 2017a).

Contrary to what European supporters have us believe, Grzimek's legacy is not uncontroversial in Africa (Homewood et al. 2009, Mbaria & Ogada 2016). Some critics regard the exclusion of humans from reserves as 'fortress conservation' and cold expropriation of local communities. Moreover, the promise of prosperity through tourism is seen as a justification that allows a powerful nature conservation lobby to inherit the controlling interests of former colonial rulers. Accordingly, national parks are perceived as miniature colonies.

In past centuries, colonialists disparaged sub-Saharan Africa as the dark continent, an epitome of human primitiveness and barbarism – a trope that often served to justify and excuse the brutal actions of the occupiers. Schürmann makes the interesting assertation that Bernhard Grzimek reinterprets the meaning of darkness in a plaque on the Ngorongoro memorial where his ashes were laid to rest on 13Mar87 (Schürmann 2017a). It reads: "It is better to light a candle than to curse the darkness." The darkness: a world without wild animals; the candle: national parks. How those should be managed, with or without excluding local people and livestock, in the context of ongoing environmental, political and societal change, is still a matter of debate. In at least one place, however, Grzimek has left a formative legacy: Rubondo Island.

Resettlements and Releases: Humans and Animals Intertwined

The developments which ultimately led to the creation of Rubondo Island National Park began in the 1950s, when the wildlife authorities of British-controlled Tanganyika realised, on the eve of its independence, that the survival prospects of the heavily hunted black rhino were alarmingly poor (Schürmann 2017b). As the last colonial official in charge, Bruce Kinloch had taken over as chief game warden of the Game Division in 1960 (Kinloch 1972). He wanted to relocate rhinos from heavily hunted areas to safe spaces. In 1961, the game warden of the Mwanza region, Peter Achard, suggested Rubondo for this purpose. Although there were no rhinoceroses there, the island seemed suitable because access for poachers was relatively difficult and there were no large predators threatening young rhinos. Thus, an island reserve appealed to the managers who struggled to ward off poachers on the mainland. At first Kinloch tried to anaesthetise rhinos with the recently invented dart gun. But, a large proportion of the animals died from incorrect dosage. Kinloch therefore hired the commercial animal trapper Thomas Carr-Hartley from Kenya. Consequently, 11 rhinos were relocated to Rubondo by boat from 1963 to 1965 (McCulloch & Achard 1965, 1969).

Introducing a large and bellicose animal prone to crop-raiding would predictably create conflict with the Banyarubondo islanders, who, apart from fishing, lived from small-scale agri- and horticulture. Thus, Achard invoked old German colonial rules stipulating the island as a forest reserve (Borner 1980: 8), a technical justification to evict the roughly 500 villagers. This, in 1964, required them to resettle on the mainland and surrounding islands – as we have described earlier (Kiwango et al. 2005).

Kinloch and Achard wanted to capture another 40 rhinos plus some other species attractive to future tourists. However, the country had by now gained independence and Kinloch gave up his post in mid-1964. His successor, Hassan Mahinda, had no interest in resettlement experiments, but instead tasked rangers with improving the protection of original wildlife habitats (Schürmann 2017b).

This prompted Achard to approach the well-known director of Zoo Frankfurt, filmmaker and conservationist Bernhard Grzimek, whose donations to wildlife causes had also garnered him prestige in the newly independent East African states. During a flight over the island in Dec64, Achard enthused Grzimek, as he later describes:

> Achard felt rather forlorn in this corner of Tanzania […]. 'All the tourists and V.I.P.s go to the Ngorongoro Crater and Serengeti,' he told me. 'Nobody comes to Mwanza for a look at my part of the world.' […] I immediately fell in love with the island. […] Three-quarters of its 135 square miles are covered with forest and the remainder consists of grass-covered hills. Above all, though, Rubondo is uninhabited.
>
> (Grzimek 1971: 12)

It seems as if Grzimek, in his excitement, got carried away about the size of the island – which actually measures 270 km² (Chapter 3), and not 350 km², into which 135 square miles would translate. However, he never cared much about the exact dimension, given that he repeats the mistake in several publications – and some of his FZS employees have followed suit (e.g. Grzimek 1966a, Kade 1967). Similarly, his praise that the "fisherfolk" had been resettled "by their energetic new government" (Grzimek 1971: 12) is misleading: "When the country was still German or British, a colonial administration would never have dared to relocate people in favour of nature and animals. They would have feared to appear hostile to the natives" (Grzimek 1966a: min 36:37–37:01). This comment belies the fact that the German colonists had been the first to empty the island of humans (Kiwango et al. 2005).

A similar get-tough sentiment about the Tanzanian powers is evident in comments about anti-poaching measures:

> The new African judge at Mwanza heard a case involving two men who had illicitly cut wood on Rubondo. 'I realize that the situation is still new to you', he told them. 'I shall therefore treat you leniently this time.' He sentenced each of them to a year's imprisonment … Under the British they would undoubtedly have got off with a fine of a few shillings.
>
> (Grzimek 1971: 12f)

As for the project itself, Grzimek soon changed strategy. Instead of emphasising Rubondo's role as a refuge for endangered species, the island was destined as a spectacular visitor attraction – serving Grzimek's conviction that only money-making tourism secures a future for wildlife. On 01Mar65, Grzimek informed the Minister for Agriculture, S.A. Maswanya, about plans to release chimpanzees and other animals attractive to visitors, suggesting that the island be turned into a national park. Its evergreen rainforest would set Rubondo apart from other reserves in East Africa and thus catalyse a larger flow of tourists to Tanzania. However, the minister rebuked these plans in a letter dated 18Jun65, because turning the island into a park would run against national interest as it would prohibit forest exploitation (Sewig 2009: 316f). Given the threat of logging, Grzimek suggested to Achard to quickly create facts, which resulted in the translocation of 6 giraffes to Rubondo in Jun65. Giraffes were by no means threatened by extinction, but they embodied tourism potential.

Since Achard fell seriously ill in 1966 and Mahinda did not initially reappoint his post in the Game Division, Grzimek enjoyed freedom of action. Accordingly, from 1967–71, a couple of dozen colobus monkeys, roan and suni antelopes were ferried to Rubondo. In addition, plans were made to embellish the island with gorillas, okapis, bongos, greater kudus and elephants (Grzimek 1971: 18). The pachyderms were to breach undergrowth, creating hiking trails for future tourists. To Grzimek's annoyance, Carr-Hartley did not provide elephants until 1972, and 4 cows at that. To build up a reproductive herd, 2 young bulls approximately 6 and 8 years old were added in May73 (Matschke 1974, Schürmann 2017a). Within this context, chimpanzees, too, were released from Jun66 to Jun69, events central to our book that are described in the next chapter. In any case, Rubondo's forests and grasslands were now home to seven types of large mammal that had not lived there before.

In early 1970, Grzimek convened a strategy meeting on Rubondo. John Owen and Myles Turner from the Tanzania National Parks Authority and

Tony Mence from the College of African Wildlife Management partici-
pated, but representatives from the Game Division were not in attendance.
In the months that followed, a new airstrip and first guest house were built.
In early 1971, however, stagnation set in because, due to the Idi Amin coup
and impending war with neighbouring Uganda, the budget of the Game
Division was reduced in favour of the military. Mahinda completely cut
funds for the Rubondo experiment, which he disliked. Consequently, FZS
would have to finance not only further species introductions, but also large
parts of the running costs of the entire island operation. These developments were met with displeasure at the Frankfurt headquarters (Schürmann
2017b). The FZS therefore binned plans for the exorbitantly expensive set-
tlement of additional animals such as gorillas, bongos and okapis.

Still in 1971, Grzimek discussed the situation with 15 local politicians
who, expecting income from tourism, voiced their favour for turning
Rubondo into a national park. However, the prospect of losing control over
the island meant even less cooperation from the Game Division (Schürmann
2017b). In early 1974, another meeting of about ten stakeholders on
Rubondo itself included Grzimek, the Regional Commissioner of Mwanza
as well as the newly appointed Director of National Parks, Derek Bryceson
(Matschke 1976) – a get-together that likely turned the tide and created a
more positive mood towards the national park idea (Gustl Anzenberger,
pers. comm.). Nevertheless, several years would pass, until finally, in Fe77,
parliament declared Rubondo as Tanzania's tenth national park – the largest island national park in Africa – and opened it to tourism.

Rubondo was now under the control of the Tanzania National Parks
Authority (commonly known by the acronym TANAPA), a parastatal
organisation established by the Tanganyika National Parks Ordinance of
1959. Designation of an area as a national park affords it the highest level of
protection. Except for recreation, local people cannot inhabit or enter these
reserves and only TANAPA or tourist lodge employees and researchers can
live inside the park.

In subsequent months, TANAPA hired more staff, overhauled trails
and the airstrip and planned for visitors. Introducing more species, however, was not on the agenda. The FZS increasingly limited itself to an advisory role. The society entrusted the young Swiss couple of Markus Borner
(1945–2020) and Monica Borner, who from 1977–84 lived intermittently
on Rubondo to assist TANAPA (Borner 1980). Markus Borner would go on
to run wider FZS Africa operations until 2012. Indeed, through the 2000s,
the FZS funded large portions of the TANAPA operational costs (Steria
Ndaga, Chief Park Warden Rubondo, pers. comm. to JNM). They withdrew their financial support in 2011, largely because by this time, and in
contrast to Grzimek's historical assertions, the island was perceived to lack
conservation value – given the fact that many of the 'attractive' animals
were not endemic. This left TANAPA with full responsibility both opera-
tionally and financially to manage the park.

Gustl Anzenberger – on Rubondo in 1974

Figure B1.1 Gustl Anzenberger watching a burrow of a carpenter bee. (Photo: Gustl Anzenberger)

Bernhard Grzimek wanted 'someone' to check on the Rubondo chimpanzees a few years after their release. He contacted Professor Konrad Lorenz at the Max Planck Institute for Behavioural Physiology in Seewiesen (Germany), where, in my fifth semester of biology studies, I lived at the time. I had been a field assistant at the Serengeti Research Institute two years prior and was excited about another opportunity to go to East Africa. Once on Rubondo, I set out to explore the vast forests – only to realise that an in-depth study of the chimpanzees was not possible in the short time at my disposal. That was the reason why instead I conducted pioneering research on carpenter bees. Many anecdotes still stick in my mind. For example, the Frankfurt Zoological Society provided the warden families with basic medication. One Sunday afternoon, a warden came to my room, yelling and clutching his belly. Not understanding Swahili, I diagnosed stomach pain. But he didn't want pills; it turned out his wife had gone into labour, and he wanted the outboard motor to ferry her to the mainland hospital. Black rhinos still existed on the island, and I bumped into one near the airstrip. Luckily the animal made no move, and I withdrew slowly enough to climb up a small tree. The rhino soon left but in the tree I discovered the nest of a Tambourin dove, which I then revisited regularly to see the two fledglings growing up. One of my sweetest memories relates to all those mornings, when at dawn I regularly would be woken by the duet of a pair of white-browed robin-chats. I can still hear their song while I lie here in my bed in Europe.

Table 1.1 Animal species introduced to Rubondo Island

Common name	Latin name	Introduction[a] (YYYY-MMM)	Origin[a]	n[a]	Sex (M:F)[a]	First offspring[b]	Status 1973[c]	Status 1978–84[b]	Remark 1978–84[b]	Status 1994[d]	Status 2002–04[e]	Status 2010s
Black or hook-lipped rhinoceros	*Diceros bicornis*	1963-Oct – 1965-Jan	Grumeti, Mwanza, Mara River[f]	16	At least 3:8	1966	25	30	Several poached	Extinct	Extinct	Extinct
Giraffe	*Giraffa camelopardalis*	1965-Jul–Nov	?	12	?	1967	16?	20	1978, 1979; poaching probable	Herds of 6–10 seen	Present	20
Chimpanzee	*Pan troglodytes*	1966-Jun, 1966-Oct, 1968-Jan, 1969-Jun	Europe (zoos, circuses)	16	7:9	1968	30	20	At least 2 males shot after attacks on humans	32 (24–40)	27–35	ca. 35
Guereza or black-and-white colobus	*Colobus guereza (aka abyssinicus)*	1967-Jun–Jul	SE slope of Mt. Meru, Arusha	20	6:19	?	40	30		At least 18	Present	20 or less
Roan antelope	*Hippotragus equinus*	1967-Oct	Maswa, south of Serengeti	5	2:3	1969	15	10	Poaching probable	Extinct	Extinct	Extinct
Suni antelope	*Neotragus moschatus*	1971	SE slope of Mt. Meru, Arusha	14	4:10	?	?	?		Present	Present	Present
Elephant	*Loxodonta africana*	1972, 1973	Arusha, Singida	6	2:4	1983	6	9	1979, 1 dead bull found	22	40	102 (72–144)[g]
African grey parrot	*Psittacus erithacus*	1990	?	37	?	?	?	?		Present	Present	30

[a] Based on information in Grzimek 1969: 34; Rodgers 1977; Borner 1988; Müller & Anzenberger 1995; details differ between sources
[b] Borner 1988
[c] Rodgers 1977; survey by 'Miombo Research Unit' and the 'College of African Wildlife Management'
[d] Müller & Anzenberger 1995: 65
[e] Moscovice 2006: 27
[f] McCulloch & Achard 1965, 1969
[g] Mwambola et al. 2016

To recap, seven types of mammals (rhinoceros, giraffe, chimpanzee, guereza, roan, suni, elephant) were released over 10 years from 1963–73. For good measure, in 1990, three dozen African grey parrots illegally captured in Cameroon and confiscated in transit were likewise set free on Rubondo (Table 1.1). While Monica Borner laments that there was very little record keeping and post-release monitoring (Borner 1988: 119), all species began to breed, some of them doubling numbers within 10–15 years. This was also true for giraffe and roan, despite scarcity of suitable habitat, i.e. open forest and savannah (ibid., 118). In fact, Peter Achard had envisioned that giraffes would keep abandoned fields and meadows as well as ridges free from tree growth (ibid., 8). Very little is known about sunis, given the cryptic habit of these dwarf antelopes. Interestingly, although elephants were supposed to create tourist paths, Monica Borner feared that they "will damage the forest when their numbers increase" and so recommended that "they should be removed again" (ibid., 119). This did not happen, and according to estimates based on dung samples, Rubondo today harbours about 100 elephants (Mwambola et al. 2016; for growth rates, see also Foley & Faust 2010). The flipside of this remarkable success are realistic concerns that larger numbers of elephants may indeed one day exceed the carrying capacity of what is after all a relatively small island with finite space, and that the ecosystem may lose its aesthetic value due to overexploitation by the pachyderms (Mwambola et al. 2016: 118).

Markus Borner – on Rubondo 1978–84

Figure B1.2 Markus Borner. (Photo: FZS)

When Prof. Grzimek convinced Dr. Julius Nyerere to make Rubondo a national park, he promised the Frankfurt Zoological Society would help to develop and maintain its infrastructure. That's why I came to Rubondo with my wife Monica and our two small children. We felt like the 'Family Robinson', shipwrecked on an uninhabited island. We learnt to bake bread in a tin pot over a fire. We truly enjoyed these times, no telephone, no hectic buzz of the modern world. In the first few weeks, we shared a small house with the newly appointed chief park warden Samwuel Maganga in Kageya – while the rangers had to make do with corrugated iron sheet boxes as their abode. Our only connection to the outside world was a short-wave radio to call to the Serengeti park. If there was a problem, we wrote a letter to Germany. It took three months to receive a reply – by which point we had forgotten our original question. The only way to reach the mainland was in a rusty diesel boat that puffed out thick smoke as it glided across the lake to Mwanza. Getting to town was quite a journey and so we only ventured out every three months. To better protect the island, we modified the wooden boats confiscated from the fishermen poachers and refashioned them with an outboard motor, which made our patrols much easier. The connection one feels of being at one with the water, the trees and the wildlife on Rubondo still makes me smile.

While plans to release okapis, bongos and gorillas had long been shelved, renewed considerations of future translocations of other endangered forest species such as the Zanzibar red colobus and the Zanzibar red duiker (Borner 1988: 119) also never materialised.

Sadly, already in the early 1980s, three species had been subject to poaching: giraffe, roan and rhinoceros – the very flagship animals translocated to protect them from such fate. Sometime after the mid-1980s, poachers exterminated Rubondo's rhinos and roan. Elephants, on the other hand, were rarely killed by hunters (JNM, pers. obs.; contrary to Mwambola et al. 2016: 118). Still, rumours have it that sometime in the 1990s, TANAPA managers were in cahoots with poachers, who killed an elephant and removed its tusks. When the plot was about to be uncovered, the corrupt officers tried to conceal their tracks by burning the carcass – an equivalent of a smoking gun that was duly noticed. More recent discoveries of dead elephants with their tusks still attached point to causes of natural death (Mwambola et al. 2016). The other introducees, giraffes, colobus monkeys, suni antelopes and grey parrots, are also still present as of today – as are the chimpanzees.

Such was the re-imagining of a wilderness that created what many indeed regard as a 'jewel' in Lake Victoria.

2 The Founder's Odyssey: Captured, Caged, Released

Figure 2.1 A chimpanzee on Rubondo in her night nest – half a century after apes were released on the island. (Photo: Paul Kivuyo, Oct19)

"The most interesting animals" introduced to Rubondo, according to Frankfurt Zoological Society's Monica Borner, are the chimpanzees, previously held captive in Europe, who "had to rehabilitate themselves with practically no help from humans" (Borner 1988: 119) (Figure 2.1). How did African apes, in the first place, end up on a foreign continent? And why were such apes transported to Africa – or, 'back' to Africa? How did these creatures, cruelly kidnapped as kids to be incarcerated for years, take to ancient, albeit entirely new, surroundings in the 'wild'? Having described the geopolitical context of the interventionist policies that embroiled Rubondo island,

DOI: 10.4324/9780367822781-2

with the subsequent eviction of humans to make room for rhino, elephant, giraffe and co., we now turn our attention to these most particular castaways: chimpanzees.

'Operation Chimpanzee': Exporting Apes to Africa

An account of the release of apes onto Rubondo accessible for non-German readers is the adaptation of a book chapter, *Menschenaffen reisten von Europa nach Afrika* (Grzimek 1969), translated as *How Europe exported apes to Africa*, first published in the United States (Grzimek 1970) and then later the United Kingdom (Grzimek 1971). Strangely, the texts call chimpanzees "monkeys", owing to the fact that the German word 'Affe' denotes *both* monkeys and apes – a crucial nuance that escaped the translators.

As part of his vision to turn Rubondo into a tourism attraction (Chapter 1), Grzimek provides different explanations for why he wanted to add chimpanzees to the island. The first related to species protection, the second to animal welfare. As for worries about the species, Grzimek formulates fears that extinction was imminent because of progress in xenotransplantation. Given a lack of human donors, the American surgeon Keith Reemtsma (1925–2000), from 05Nov63 to 10Feb64, performed 13 transplants of kidneys from chimpanzees to humans. Most recipients died within a few weeks, but one woman survived for 9 months, aided by the new immunosuppressants, and was even able to continue her work as a teacher during that time (Reemtsma 1995). These trials alarmed Grzimek, as he envisioned an enormous demand for chimpanzees for medical purposes to transplant their organs to humans: "Scientists are already working on this problem throughout the world and particularly in the United States, where 30,000 people die of kidney disease alone in a single year." Hence, once the techniques of kidney grafting are developed, "it may be only a year before all chimpanzees of the appropriate blood-group are exterminated." Thus: "It would be a good plan to establish a few island sanctuaries for them here and there" (Grzimek 1971: 18). Today we know that Grzimek's fears proved unfounded, because further experiments failed and dialysis machines soon eased the pressure to procure donor kidneys (Reemtsma 1995).

In addition, Grzimek justified his focus on chimpanzees with animal-welfare considerations:

> Most zoos only want young animals and are not equipped to keep full-grown, sexually mature chimpanzees. They liked to show off the amusing mischievous youngsters and didn't know what to do with them once they had grown up. [...] It distressed me to see such intelligent and active creatures shut up singly or in pairs in rectangular zoo cages, so I was delighted to be able to give them their freedom. [...] Incarcerating anthropoid apes by themselves was a brutal thing to do – at least as bad as keeping a convict in solitary confinement.
>
> (Grzimek 1971: 16)

While these are reasonable sentiments, there is a certain degree of hypocrisy in these worries, given that Grzimek doesn't object to the trapping, incarceration and exploitation of chimpanzees for the entertainment industry in the first place. Housing them in larger groups instead of alone or in pairs is a rather cosmetic mitigation of this problem – not to mention that only a couple of apes were set free, while many were left behind.

Indeed, Grzimek's ideas about animal welfare come across as quite chequered. In one of his writings, he compares his role in the zoo with that of "a prison director", who "is often blinded and dulled to what is really going on there. A person can even get used to a concentration camp" (Grzimek 1974: 240). However, he then labels this a flawed comparison, suggesting instead that "progressive" zoos provide such good care for their charges that the animals would voluntarily live there, as they are fed, safe from predators and housed in mixed-sex groups where they can reproduce. Similarly contradicting is Grzimek's proposal to reduce the number of species kept in zoos so that those remaining have more space, while he still speaks favourably of circuses – where animals are kept in tiny cages – and condones the practice of drilling zoo-chimpanzees to perform tricks for show purposes, claiming that "such tasks are great fun for zoo animals" (Grzimek 1988: 483). Given all this, it seems logical that Grzimek tried to eliminate the wording of zoo animals being held 'in captivity' ('in Gefangenschaft'), systematically substituting this term with the euphemism 'in human care' ('in Menschenobhut') (Gustl Anzenberger, pers. comm).

To obtain the desired apes for Rubondo, Grzimek approached various parties to ascertain whether they might sell chimpanzees for a reasonable price, including zoos in Europe and the United States as well as two primate research centres and the Schweizer National Circus Gebrüder Knie. Knowing only too well that young apes were visitor magnets for zoos, while adults constituted a burden, he remarks in his letter: "I believe that in many cases the owners would be quite happy to get rid of these animals" (cit. in Schürmann 2017b: 281).

Grzimek also gives reasons why, unlike what was done with the rhinoceroses, he did not opt to translocate wild chimpanzees. In response to questions about him "bringing chimpanzees to Africa from Europe" when he "could have captured them in East Africa", he explains: "Nobody captures full-grown anthropoid apes in Africa – they are too dangerous, too strong and quick-witted" (Grzimek 1971: 16). This rationale differs from what he says later in the TV episode about the releases, where he (incorrectly) maintains that there are no chimpanzees in East Africa. Still, here he also alludes to the need to only release animals that are old enough: "You might wonder why we bring zoo chimpanzees here to Africa. Well, there are no great apes in this area. We'd have to catch them in West or Central Africa, and only chimpanzee children are caught there. Great apes are too strong and too dangerous. [...] After all, we cannot release a flock of small children alone in the wilderness. They would be hopelessly lost. And that's why we have to resort to large zoo chimpanzees" (Grzimek 1966a: min

00:18:22–00:18:56). Again, it is disconcerting that Grzimek's trepidations for a species overall do not translate into concern for individuals, such as baby apes who are subjected to suffering the massacre of their mothers and community members while being caught by animal traders.

In response to Grzimek's circular, offers were soon received, but mainly for males. Grzimek approached the Ludwig Ruhe company, one of the largest animal dealerships globally, headquartered in Alfeld in Northern Germany, to source additional females, but without success. Because of this problem and because Achard, his main partner on the ground in Rubondo, suffered a stroke, Grzimek postponed the operation until the following year. Finally, in the spring of 1966, Grzimek had assembled a first batch of chimpanzees from European contacts: 3 males, 8 females. Another 6 animals would be sourced later, bringing the total to 7 males and 10 females.

What we know about the provenance and subsequent shipment of these charges to Africa is summarised in Table 2.1. While we discuss the interesting question about their geographic origin in a later section, the 17 apes had somehow ended up in at least seven European nations (Sweden, Denmark, the Netherlands, France, Germany, Austria, Switzerland). Before being bound for Rubondo, they had been kept between 3.5 months and 9 years in captivity, including zoos (15 animals), circuses (4 animals) or with animal traders (3 animals), often changing hands between these owners.

Some housing conditions included solitary periods, at times under severe spatial restrictions. All but one of the chimpanzees (who was kept with orangutans) were housed with at least one other chimpanzee, and at times more conspecifics. One or two chimpanzees spent time in human homes as infants. This includes Kathrin (also named Caroline) (Figure 2.2), who served as a pet in the Ivory Coast residence of the Swiss biologist Hans Huggel. At least three chimpanzees (nos. 15–17, cf. Table 2.1) had undergone the typical 'career' of circus apes, who, once they had outgrown their infantile 'cuteness' and thus their utility as being attractive for spectators, were sent to a zoo.

Female Jette (no. 15, cf. Table 2.1), born around 1960, gave birth to two infants in 1967 and 1968 without raising them, indicating a time of first birth at age 7 (such early primiparity can occur in captivity; Littleton 2005, Carlsen & de Jongh 2014). Upon the time of their travel to Africa, 9 of the chimpanzees were familiar with at least one other captive comrade (cf. Table 2.1: nos. 3 and 4; 5 and 6 [probably]; 10 and 11; 15, 16 and 17), but the others were not.

When destined to be released, the animals were ca. 4–11 years old (overall median 8.8; female mean 8.1, range 4–11; male mean 8.1, range 6–10), i.e. juvenile to subadult. Some contradictions about ascribed ages could not be resolved. For example, Grzimek (1970: 13f) states about the animals sent out as the first cohort: "The seven females were full-grown, only one of the males – the muscular Robert – was full-grown, the second being immature, and the third little more than a baby." However, our compilation (cf. Table 2.1) indicates that, when sent to Africa, no chimpanzee had reached

Table 2.1 Provenance of the founder population of Rubondo chimpanzees

No.	Release date on Rubondo	Name (studbook number)	Sex	Origin	Date of birth	Captivity (place, duration)	Captivity (housing condition)	Acquisition for Rubondo release	Approximate age upon release (yrs)
1	1966-Jun-23	Simba	F	Africa	1956–1957	Since 1960 at *Furuviksparken*, Furuvik, Sweden (6 yrs)	Furuvik: housed with other chimpanzees; no offspring; no serious illness recorded	Donated jointly by Gottfried Fridh of Stockholm and *Furuviksparken*	9
2	1966-Jun-23	NR	F	Sierra Leone	NR	Directly from Africa to animal trader *Jabria*, Harderwijk, Netherlands (3.5 months)	Kept with 2 similar-aged chimpanzees; weight on departure to Rubondo 17 kg; no serious illness recorded	Bought from *Jabria*	4
3	[Died 1966-Jun-06 in transit from heat stroke after stopover at Port Said]	Joséphine	F	Probably Guinea	1957	*Parc zoologique et botanique de Mulhouse*, Mulhouse, France (when 18 months old)	Joséphine and Schilla arrived 1964 and were kept in improvised enclosure ("Notunterkunft") without outdoor access; no serious illness recorded; weak physical and psychological condition	Presumably exchanged	[8]
4	1966-Jun-23	Schilla	F	Probably Guinea	1957–1958	*Parc zoologique et botanique de Mulhouse*, Mulhouse, France	ditto	Exchanged	8–9
5	1966-Jun-23	Lillemor (10760)	F	Africa	ca. 1960	Garden Society of Gothenburg (*Trädgårdsföreningen*), Sweden; ca. 1962 to *Borås Djurpark* (aka *Berggren Zoo Park*), Sweden	Probably kept with other chimpanzees	Gift from *Borås Djurpark*	6

(Continued)

Table 2.1 Continued

No.	Release date on Rubondo	Name (studbook number)	Sex	Origin	Date of birth	Captivity (place, duration)	Captivity (housing condition)	Acquisition for Rubondo release	Approximate age upon release (yrs)
6	1966-Jun-23	Ricky (10769)	M	Africa	ca. 1960	ditto	Kept with other chimpanzees in small cage; weak physical and psychological condition	ditto	6
7	1966-Jun-23	Lola	F	Probably Africa	ca. 1955	*København Zoo*, Denmark, handed over by officers of ship MS Norden (1957-Nov-20), probably directly from Africa	ca. 2 yrs old on arrival; always kept with other chimpanzees	Gift from *København Zoo*	11
8	1966-Jun-23	Bastian	M	Africa	ca. 1962	Bought from animal trader Krag (Copenhagen), delivered to *København Zoo*, Denmark (1963-Jun-25)	ca. 1 yr old upon arrival; 1 month in solitary confinement, afterwards kept with 2 orangutans	ditto	4
9	1966-Jun-23	Kathrin (Vienna, Austria) resp. Caroline (Basel, Switzerland)	F	Probably Ivory Coast	ca. 1956-Jun	From residence of Prof. Hans Huggel, Ivory Coast (1 yr), sent directly to *Zoo Basel*, Switzerland (1958-Apr-25); *Tiergarten Schönbrunn*, Vienna, Austria (1963-Jul-29); Frankfurt zoo (1966-May-11)	Huggel-residence Ivory Coast: kept with human children and other non-ape primates; Basel: kept with slightly older female chimpanzee and similar-aged orangutans and gorillas; Vienna: at first together with chimpanzee male, then separated	Bought from *Tiergarten Schönbrunn*	10

10	1966-Jun-23	Kathy	F	Probably Africa	ca. 1958–1959	Taken over from a Danish captain named Hansen; *Zoo Gelsenkirchen*, Germany (1961-Jun-15 – Dec-17); (Zoo?) Dortmund, Germany (1961-Dec-17 – 1965-Nov-18); afterwards *Zoo Duisburg*, Germany	Gelsenkirchen, Dortmund: kept with other chimpanzees; Dortmund (Duisburg?): Kathy and Robert kept together	Bought from animal trading zoo *Tierpark Schloss Krechting* (Tierhandelszoo Konrad Müller)	5–7
11	1966-Jun-23	Robert	M	Probably Africa	1958–1959	ditto	ditto	ditto	5–7 ("fully grown" according to Grzimek 1969, 1971); (likely shot dead after attacking game wardens; Borner 1985: 152)
12	1966-Oct-19	Jimmy (10676; name=Demo)	M	Africa	ca. 1958	First in circus, then in *Zoo Herford*, Germany (from 1962); 1966-Feb-01 *Zoo Frankfurt*, Germany	Herford: kept with 2 female chimpanzees; blind left eye due to corneal injury	From *Zoo Frankfurt*	7 or 8; (shot dead 1968-Oct after attacking volunteer and game wardens; Grzimek 1970: 37)
13	1968-Jan	Jimmy [sic] (10693)	M	Africa	ca. 1958	*Tierpark Lübbecke, Germany; Zoo Frankfurt 1967-Oct-27*	NR	ditto	10
14	1969-Jun-30	Peggy	F	NR	ca. 1960	Acquired from animal trader Krag (Copenhagen) by *Zoo Wuppertal*, Germany, in 1961 (ca. 8–9 yrs in captivity)	NR	Gift from *Zoo Wuppertal*	9

(Continued)

Table 2.1 Continued

No.	Release date on Rubondo	Name (studbook number)	Sex	Origin	Date of birth	Captivity (place, duration)	Captivity (housing condition)	Acquisition for Rubondo release	Approximate age upon release (yrs)
15	1969-Jun-30	Jette	F	NR	1960	Donated to *Zoo Wuppertal*, Germany, by animal trainer Konzelmann of *Tierpark Hagenbeck* (Hamburg) once found too big for circus performance (≥8 yrs in captivity)	Zoo: Jette, Bobby, Jeff kept as one group; Jette gave birth in 1967 and 1968, did not raise her infants	ditto	9
16	1969-Jun-30	Bobby	M	NR	1959–1960	ditto	ditto	ditto	9–10
17	1969-Jun-30	Jeff	M	NR	1959–1960	ditto	ditto	ditto	9–10

Note: NR = no record; ditto = information from table cell above applies. Dates given as YYYY-MMM-DD. Data compiled from: (i) Grzimek 1966a, 1969; (ii) records obtained by Gustl Anzenberger in 1973 from Zoologische Gesellschaft Frankfurt (Müller & Anzenberger 1995); (iii) Schürmann 2017b; Table 1; (iv) archival records of Felix Schürmann: Institut für Stadtgeschichte, Frankfurt am Main, Germany (Zoologischer Garten 197 and 212); Zoologische Gesellschaft Frankfurt (Afrika, 00001970, Rubondo 3, Keller Regal 01021; 00001976, Rubondo 1, Keller Regal 01019); (v) European chimpanzee studbook (Carlsen & de Jongh 2014). (Modified after Msindai et al. 2021)

Figure 2.2 Chimpanzee Caroline (pictured with keeper Carl Stemmler), kept at Zoo Basel, was one of the animals acquired by Grzimek to be shipped to Rubondo. (Photo: Zoologischer Garten Basel, Switzerland)

full adulthood (i.e. 13–15 years of age; cf. Havercamp et al. 2019). Still, if not already reproductive, these animals would have soon reached that age, given that in captivity in Europe, the youngest (confirmed) age of reproduction for a female is 6 years and 9 months, and for a male 6 years and 1 month (Carlsen & de Jongh 2014).

The animal dealership Ruhe in Alsfeld, by the spring of 1966, sent the first assemblage of 11 chimpanzees by truck to Antwerp in the Netherlands (Schürmann 2017b: 282). Here they were put on the Eibe Oldendorff, a cargo steamship of the Deutsche Afrika-Linien (DAL). Grzimek was full of praise about the transport conditions:

> The crates in which they were housed were capacious and easy to keep clean. The animals had to travel singly or in pairs because they did not know each other and would soon have come to blows if herded together at close quarters. [...] Our head keeper Gerhard Podolczak [did] accompany them on the freighter. Captain K.W. Wehlitz took a keen interest in the hirsute passengers on his freighter. He had sturdy wooden sheds built on deck so that the animals need not go below. [...] The chimpanzee villa on deck, 40 feet long and 6 feet wide, was covered with two layers of tarpaulin by the ship's carpenters and even fitted with internal electric lighting.
> (Grzimek 1971: 14)

On 21May66, the ship set sail for Mombasa. The chimpanzees remained at sea for over three weeks:

> The ship only put in twice, once at Port Said on May 1st in order to load 10 tons of oranges and a second time at Djibouti in order to unload

them. Despite air-conditioning and sprinklers, eleven days in the Red Sea proved to be rather more than the monkeys and their keeper could stand.

(Grzimek 1971: 14)

When Grzimek alludes to adverse conditions, it reflects his usual strategy to add suspense and grit to his stories. However, in his public accounts, he never reveals the death of the female chimpanzee Joséphine, who, between Port Said in Egypt and Djibouti, suffered a fatal heat stroke on 06Jun66 (Müller & Anzenberger 1995: 12). Consequently, because the ship left with 11 chimpanzees and because only 10 survived, Grzimek's accounts – and subsequent reports of others – are riddled with conflicting information about numbers and sexes of the animals that were ultimately set free (see below).

In the meantime, John Owen, director of the National Parks Authority, had informed the aspiring primatologist Jane Goodall about the project. Grzimek knew that his own actions did not compare favourably with Goodall's long-term observations of wild chimpanzees at Gombe in Tanzania (van Lawick-Goodall 1971). Indeed, Goodall would later state that "this was something that I strongly disapproved of. There was no real plan" (cit. in Sewig 2009: 318). Grzimek reprimanded Owen and resorted to a new terminology through which he sought to protect him against criticism in case of failure: "It is an experiment and I do not want to have too much publicity even among scientists" (cit. in Schürmann 2017b: 283).

The plan to transfer the surviving animals onto a train in Mombasa fell through because the port was overcrowded. The ship instead headed for Dar es Salaam where the crates were unloaded on 12Jun66, not without attracting attention. "Loudly, each chimpanzee cage that was hoisted ashore was cheered and applauded" by African dock workers. "Even the crane operator hoisted the cage high enough each time so that he could get a good look at the chimps" (cit. in Schürmann 2017b: 283). A newspaper printed what Grzimek labelled a "ridiculous report", that the animals from European zoos "were accustomed to nothing but the best Russian tea". Consequently, the main problem would be "how to convert them to drinking plain water in the wild. I don't know which sailor sold this nonsense to the African reporter in Dar, but the same picture and report were reprinted in every German news-paper" (Grzimek 1971: 15) (Figure 2.3).

Grzimek had dispatched two trucks and two off-road vehicles to the coast, along with a crew of friends and experts. Gerhard Podolczak continued the journey, now joined by a young Scot, Sinclair Dunnett of a British volunteer organisation, who stood in for Peter Achard. They were joined by the German biologists Fritz Walther and Ulrich Trappe, who were conducting research in the Serengeti, and Gordon Harvey, former head gamekeeper of the Serengeti, who had been asked by Grzimek to film the operation for his TV programme. Grzimek himself had flown by plane to Tanzania

Chimps with a problem

TEN pampered chimpanzees who refuse to drink anything but peppermint, herb or best-grade Indian tea, have arrived in Dar es Salaam to be returned to the wilds on a Lake Victoria island where famed German zoologist and author, Prof. Bernhard Grzimeck, hopes to start a new national park.

The chimps, donated by nine zoos in the Scandinavian countries, Germany, and France, arrived in Dar es Salaam aboard the Elbe Odondorff, which is on charter to the German East Africa Line.

Their diet on the long sea voyage — they were sick for half a day when the vessel met heavy weather — has been fruit, white bread, rice and lukewarm tea served without milk or sugar. They absolutely refused to drink water.

They will be shipped to Rubondo, a 30 by 25-mile island in the south-west of Lake Victoria, as the first consignment of animals for a sanctuary which Prof. Grzimeck hopes one day will become a national park.

The first thing the three males and seven females will have to be taught is how to fend for themselves, for they certainly will not get lukewarm tea on uninhabited Rubondo.

"Spoon-fed" and raised in European Zoos, another problem they are meeting is the tropical climate.

The chimps became great favourites with the sailors on the long voyage. One is known as "Little Berliner", another specialises in undoing shoelaces and then neatly re-tying them, while a third holds out a piece of straw to any passer-by to scratch his ear.

Figure 2.3 Chimpanzees arriving at Dar es Salaam port make front page news. (*The Standard*, Tanzania, 17Jun66)

and met the chimpanzee convoy in Mwanza at the southern shores of Lake Victoria on 15Jun66.

Trappe and Walther embarked on a 7-hour boat trip, ferrying food, tents, beds and crates to Rubondo. Grzimek, Dunnett, Harvey and workers of the wildlife division drove 200 km "along circuitous and incredibly poor roads" (Grzimek 1971: 17) to a coastal village from where the crossing to Rubondo was only two hours.

The release took place the next morning, on 23Jun66 (Figure 2.4), in the eastern bay between Nyakutukula and Lukaya, a site chosen because of its shallow waters and distance from the park headquarters.

> At dawn, while the lake was still calm, we transported the animals in two batches twenty minutes farther along the coast to a clearing beside the shore where the beach, which consisted of smooth stones, shelved gradually. The heavy crates were carried ashore on our shoulders and placed in a row close to the water's edge. The animals were given a last dose of Resochin syrup against malaria, a disease to which chimpanzees are susceptible. [...] We up-ended the crates containing the most dangerous inmates so that their sliding hatches could be pulled open from the water with ropes.

Because apes have no innate ability to swim,

> it was improbable that they would pursue us into the water, so we should easily be able to evade them by wading in. A few hundred yards away in either direction stood the deserted banana groves of the original inhabitants. Here, as at many other points on the island, the animals would be able to obtain food until they had learnt to find their own in the forest proper.
>
> (Grzimek 1971: 20)

While some animals were reluctant to leave the crates, two females

> simply made a bee-line across the clearing and vanished into the forest as though they had always lived there. Others went to neighbouring crates which still contained chimpanzees, stuck their fingers inside, felt the faces of the inmates and exchanged kisses through the bars. [...] The half-grown immature male and the smallest chimpanzee of all, still a baby, made straight for us and clung to our legs. The baby climbed up me, kissed me and begged to be cuddled. It ran after us wherever we went, mostly on two legs, which looked quite comical.
>
> (Grzimek 1971: 22)

After placing piles of bananas, apples and bread near the crates, the crew left. They visited the release site again late that afternoon by motor-boat, on

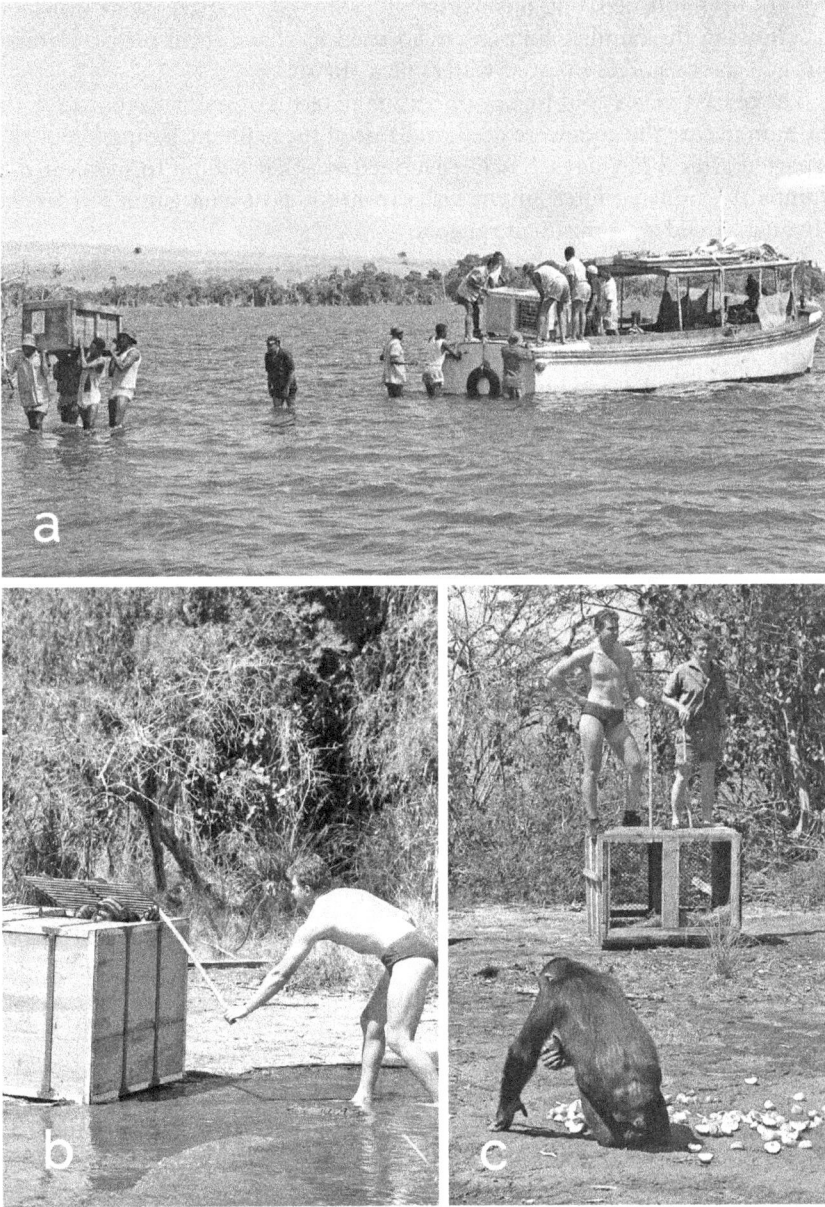

Figure 2.4 Releasing chimpanzees onto Rubondo. (a) Crates are taken to shore. (b) Chimpanzee crate opened by Sinclair Dunnett. (c) One of the first chimpanzees set free, with Sinclair Dunnett (R.) on top an empty crate. (Photos: Bernhard Grzimek/Okapia)

foot during the next morning and again by boat in the afternoon. The food had not been touched, but, unsurprisingly, some chimpanzees were insistent in following the familiar humans, who tried to shake them off by wading through the crocodile-infested waters near the shore.

The release was a cold-turkey operation in terms of severing contact with the human care the apes were used to. Most of the team, including Grzimek, departed after a few days. Sinclair Dunnett was left behind to monitor the chimpanzees during intermittent visits from his post as a game warden in Mwanza, aided by Tanzanian rangers.

On 19Oct66, another chimpanzee arrived from Frankfurt zoo by plane, to increase the number of fertile males. This was the one-eyed Jimmy, who, while in captivity, had proven dangerous and severely wounded his keeper. "After some safety precautions, we released him in a light drizzle. He looked around once and probably thought the whole thing was a mistake, then he galloped across the clearing and disappeared between lianas and thick bushes" (Kade 1967: 31). Yet another male, likewise called Jimmy, was flown over from Europe in Jan68. Ultimately, one and a half years later, on 30Jun69, 2 more females and 2 more males arrived, bringing the total number of released chimpanzees to 16.

How would they fare on the island?

'Increasingly Timid': Early Years Post Release

The fate of the chimpanzees during their initial years on Rubondo is only sketchily documented – not least because close-up encounters with the apes could lead to injury. The personnel on Rubondo therefore faced the dilemma that searching for and observing the releasees conflicted with the aim to discourage them from approaching humans. In any case, what we have are scattered remarks in a few publications, in unpublished reports and correspondence stored in FZS archives (Table 2.2).

Over the first decade, from 1966 to 1975, Grzimek tasked a succession of German forester volunteers to keep an eye on the released animals in general, while also assisting with building up infrastructure in the reserve (Ulrich Kade, 1 year; Hans Sönsken, 2 months; Wolfgang Brockmann, 8 months; Wolfgang Matschke, 2 years).

Initially, rice, bananas and bread were provided to the chimpanzees, based on the assumption that they were as yet unable to forage independently. However, it is doubtful that these rations were substantial, not least because the apes exploited cultivated crops that still grew on abandoned fields (banana, papaya) as well as figs and the pith of the papyrus plant (Kade 1967). In fact, the provisioning was ceased 2 months after the first ten chimpanzees were released.

In an article dating to about Feb67, Ulrich Kade states that, 11 months post release, he never observed the chimpanzees constructing night nests, and that they instead slept in tree forks, including a very large fig tree

Table 2.2 Monitoring of Rubondo chimpanzees during early stages post release. Dates given as YYYY-MMM-DD

Date	Event (x:y = m̄ale:female)
1966-Mar-23	**FIRST RELEASE: 3:7 (north of Lukaya)**
1966-Apr, early	2:0 youngsters 9 km from release location in abandoned banana plantation
1966-May, end	Food provisioning stopped because of sufficient wild foraging (banana sprouts, leaves, wild fruit, seeds); animals sometimes eat faeces
1966-Jul-10	1:0 with ticks and in bad condition in camp
1966-Jul-10	0:3 at release location
1966-Jul-29	1:0 from mid July OK again
	0:2 at release location
1966-Aug-13	1:0 adult ca. 10 km from release location
1966-Sep	2:0 youngsters in good condition
	Chimpanzees eat lots of figs
1966-Oct-06	2:0 youngsters often seen, displaying more vigilance towards humans
1966-Oct19	**SECOND RELEASE: 1:0 (one-eyed Jimmy) (north of Kageya)**
1966-Dec-05	Traces of chimpanzees up to 15 km from release location
	2:0 youngsters flee when boat approaches, otherwise still very trusting
1967-Jan-13	Female chimpanzees at lakeshore
1967-Feb	0:4 adults seen, later also 1:0 (Robert)
	No chimpanzee yet observed to build night nests, 11 months post release
1967-Mar-05	1:0 (one-eyed Jimmy) seen again for first time (with 0:2 or 0:3)
	2:0 youngsters still at same location
1967-May-17	2:0 youngsters increasingly timid
1967-Jun-10	2:0 youngster now in forest, no longer seen
1967-Sep-14	1:2 chimpanzees (Robert and 0:2 pregnant) "making a mess" in camp 2; Ulrich Kade is bitten
1967-Oct	1:2 seen twice, moving long distances each day, not visiting camp
1968-Jan	**THIRD RELEASE: 1:0 (north of Kageya)**
1968-Feb-21–23	2 newborns seen with their mothers
1968-Feb-16	2nd visit of 1:0 Robert and 0:2 in camp, deterred with blank shots
1968-Mar	3 chimpanzees at camp 2
1968-Apr	6 chimpanzees at camp 3
1968-Oct-10	1:0 (one-eyed Jimmy) shot after injuring two rangers
1969-Feb-22	2 chimpanzees seen
1968-Mar-27–28	3 chimpanzees at camp 2
1969-Jun-30	**FOURTH RELEASE: 2:2 (north of Lukaya)**
1969-Jun-14	1:3 at Mlaga
1969-Jul-23	0:1 (Jette) at Lukaya
1969-Jul-26	4 chimpanzees (1 young) at Kageya
1969-Aug-24	6 chimpanzees at Kageya
1969-Oct	Rangers often see small groups of chimpanzees with 2 young between Mlaga and Lukaya
1969-Nov-23	5 chimpanzees between Mlaga and Kageya
1970-Oct	Chimpanzee seen in camp with vervet monkey baby for 5 hours, playing, caressing, petting etc.; unknown, if accidental find or result of hunting; subsequent development also unknown
1970-Nov-04	Hans Sönksen bitten several times by a chimpanzee

Note: Edited version of a compilation of FZS records and letters in Müller & Anzenberger 1995: 35

(Kade 1967). However, after around a year, they went on to build sleeping platforms (Borner 1985). The chimpanzees slowly avoided contact with humans, and some travelled increasingly further from the release location, as far as 10 km after 5 months and 15 km after 7 months. Of the 11 animals released in 1966, all were still alive in Apr67 (Kade 1967).

In terms of sociality, two young males kept to themselves during the first year. About 6 months after his release, one-eyed Jimmy was seen to travel in the company of females. Similarly, adult male Robert was observed with two females, presumed pregnant, in Sep67, about 15 months post release. Two new-borns were recorded in Feb68, about 2 years post release.

Subsequently, the chimpanzees began to range widely across the island. Unlike what happened to rhinoceroses, giraffes and roan, there is only a single mention that "chimpanzee poaching (for bones for witchcraft?) has occurred" (Rodgers et al. 1977: 121).

While first births indicated that the 'experiment' had succeeded, deaths certainly occurred – resulting from the fact that some of the ex-captives continued to pose a threat to humans. Grzimek had envisioned as such: "I was quite aware that our project had its dangers" (Grzimek 1971: 11), going on to elaborate:

> Because of its shorter legs, a male chimpanzee only comes up to a man's chest when standing erect, but it has muscles twice as powerful, fangs not much inferior to those of a leopard, and a brain more highly developed than that of any living creature on earth with the exception of man. An angry wolf or leopard may sink its teeth into a broom-handle; an angry chimpanzee always bites to draw blood.
>
> (Grzimek 1971: 11)

While Grzimek certainly relishes the opportunity to add thrill to his stories, chimpanzees can truly inflict severe injury. Captive apes often injure humans because they have lost fear, lest their caretakers wield bats and water hoses: "Zoo chimpanzees [...] have no fear of man but are not necessarily tame or friendly" (Grzimek 1971: 19). Another likely trigger for aggression is frustration and emotional upset caused by long-term confinement, given that habituated wild-living apes almost never attack researchers. Apart from such factors related to captivity, wild chimpanzee males in particular are known to use force to obtain resources, and this inclination will also apply to a coveted food in the possession of humans (e.g. Goodall 1986).

Indeed, there were several assaults on zookeepers before the releases took place, as well as on FZS personnel and rangers on Rubondo itself. These attacks led to the shooting of at least one male and probably two, although rumours and statements persist of up to three being shot. In addition, the chimpanzees, during the early years, broke into houses, spurring anecdotes, some rather amusing, albeit perhaps not entirely trustworthy, of the apes behaving in anthropomorphic ways. Here is a compilation of the records:

- May66: "One of the male chimpanzees attacked Horst Klose, our experienced and athletic keeper of anthropoid apes at Frankfurt Zoo, and escaped. The animal bit through his shoe, severing a toe, and lacerated his thigh and hand. Surgeons at a nearby hospital spent two-and-a-half hours sewing him up, and he remained there for over four months" (Grzimek 1971: 12; because of this incident, one-eyed Jimmy was dispatched to Rubondo in Oct66).
- May–Jun66, about chimpanzees on board the Africa-bound ship: "Some of them were very vicious and [...] tried to grab and bite anyone they could reach through the bars" (Grzimek 1971: 19).
- Sep67, 15 months post release: "The male chimpanzee Robert turned up at the Game Wardens' new second camp with two pregnant females. The animals tore up sacks of sugar and grain, scattered things about and behaved in a thoroughly refractory manner. Ulrich Kade was bitten in the hand while trying to drive them away. The game wardens repelled later invasions of the camp by firing shots in the air" (Grzimek 1971: 36).
- 10Oct68, 24 months post release: "One-eyed Jimmy broke into the new camp and immediately attacked Game Warden Lucas Seremunda. Determined to keep the animal away from his children, the man refrained from running into the house and made for the lake-shore, but Jimmy caught him before he reached the water. In a rage, the chimpanzee bit him on both hands and would not let go until hit on the back with a stick by another game warden. Then it ran off. Six days later it reappeared and attacked Game Warden Daniel Obaha, who was sitting outside camp reading a book with a rifle at his side. Obaha tried to run inside a hut when he saw Jimmy heading for him, but the furious animal tore a piece of flesh off his leg and forced its way into the hut as well. It then closed the door from inside. In the fierce struggle which followed Obaha lost the little finger of one hand and had the other badly mauled. He fortunately managed to wrench the door open. One of the porters picked up Obaha's rifle, which was laying on the ground, and shot the chimpanzee dead" (Grzimek 1971: 36f).
- Ca. 1968: "According to rumours, at least one other adult male was shot because of similar behaviour" (Borner 1985: 152); resp. affirmative: "It is true that other chimpanzees were secretly killed by fearful rangers" (Borner 1985: 8) and "At least two males were shot" (Borner 1988: 122); similarly, "Zwei der schlimmsten Übeltäter mußten sogar

geschossen werden" ("Two of the worst offenders had to be shot") (Matschke 1976: 25); as well as: "Three animals were shot for attacking humans" (Rodgers 1977: 121).

- 04Nov70, a couple of years post release: "Hans Sönksen several times bitten by a chimpanzee" (Müller & Anzenberger 1995: 35).
- 1973: "The Chimpanzees are migrating throughout the island. [...] Favourable are the southwestern parts of Lukukuru and the Kamea areas. No serious incidents were reported this year. They entered camps several times but seemed to be more shy than before. Only the guest-house suffered some damage when Chimps entered it through the windows when people were away" (Matschke 1974).
- 1978–84, more than a decade post release: "We were at the house one day, and we saw [a female and her young]. I took out my camera which had a large telephoto lens attached to take pictures of the chimpanzees from inside the house. [...] One chimpanzee came up to the window grabbed my lens and pulled my arm through the window frame, [she] bit my arm and I still have the bite mark to prove it" (Markus Borner, pers. comm. to JNM; see also Schneider 2017).
- 1978–84: some introduced animals continued to be quite aggressive and unafraid of humans, at times breaking into houses, so that the Borner family had to board up their windows (Monica Borner, pers. comm. to JNM).
- 1978–84: "Another time we had left the island, upon our return we found the house in disarray, wine bottles broken, flour scattered all over the beds, bed sheets rumpled as if the chimpanzees had slept in the bed. We think they also washed up some of the plates and cups they had used, as you know some of these chimpanzees would have performed perhaps in tea drinking parties whilst in captivity [...]. And we think they drank some bottles of wine" (Markus Borner, pers. comm. to JNM).
- 1980/1990s, a decade or two post release: "Shadows passing by the windows and chimpanzees being seen in the evening, peering into ranger's house" (TANAPA managers, pers. comm. to JNM).
- 1980/1990s, a decade or two post release: "A female chimpanzee seen with a blanket" (TANAPA managers, pers. comm. to JNM).
- 1998–2001 or later, decades post release: "Chimpanzees broke into the food store and took and broke bottles of liquor and ate some food. Creating a lot of mess" (employees of Flycatcher Tourist Lodge, pers. comm. to JNM).

While anecdotal reports indicate that some founders survived for decades (Matsumoto-Oda 2000: 17), incidences of assaults and break-ins were bound to sooner or later become rarer if and when the ex-captives passed away. However, at least theoretically, some founders may *still* be alive and continue to do so until 2030, considering that chimpanzees can reach an age of 70 years (Havercamp et al. 2019)

Monica Borner – on Rubondo 1978–84

Figure B2.1 Monica Borner with one of her children. (Photo: Monica Borner)

The Frankfurt Zoological Society had employed me, together with my husband Markus Borner, to help establish the newly declared Rubondo National Park. In terms of the most memorable experience, I guess it was living with the introduced animals. The chimpanzees were still aggressive and unafraid of humans. One particular time we had left the island, and upon returning to our house, we found it in complete disarray. The food store had been broken into, there was flour scattered all over the place, and some of the wine bottles had been smashed. It seemed the apes had broken in, possibly in search of food and had drunk some of the wine. The beds had been slept in too, as the sheets were all crumpled. In the end we had to board up our windows to stop them entering our home.

Research on Rubondo: Primatologists on the Prowl

During the first three decades post release, the FZS arranged for follow-up surveys on the chimpanzees, with outputs largely aiming to guide park management decisions. Only from the mid-1990s onwards did blue sky research into the behaviour and ecology of the released apes take off (Table 2.3) – often not directly associated with the FZS and typically with limited continuity. Here we sketch out the main trajectories, while summarizing important findings in later chapters.

As has been mentioned (Chapter 1), back in 1974, the German ethologist Gustl Anzenberger (University of Munich and Max Planck Institute for Behavioral Physiology, Germany), spent 3 months on Rubondo.

Table 2.3 Post-release surveys and subsequent research of the Rubondo apes

Period	Name	Institution, position (activity)	Important output
1966	Bernhard Grzimek	FZS (project initiator)	Grzimek 1966a, 1966b, 1966c, 1967, 1969, 1970, 1971, 1988
1966-Jun–mid 1967	Sinclair Dunnett	Game warden, Mwanza (irregular surveys)	
1966-Jul–1968-Mar	Ulrich Kade	FZS (German forester volunteer, surveys)	Kade 1967
1969-Feb–1971-Jan	Hans Sönksen	ditto	
1971-Jan–1972-Sep	Wolfgang Brockmann	ditto	
1972-Sep–1975-ca Aug	Wolfgang Matschke	ditto; Deutscher Entwicklungsdienst (DED)	Matschke 1974, 1976
1974-Jan-10–Apr-01	Gustl Anzenberger	FZS (surveys)	Anzenberger 1977
1978–1984	Monica Borner, Marcus Borner	FZS (tasked to develop newly declared Rubondo Island National Park)	Borner 1980, 1985, 1988
1994-Oct–Nov	Guido Müller	Univ. of Zurich, Switzerland (socioecological research)	Müller 1995, Müller & Anzenberger 1995
1996–2002	Paula Robinson, Johan Robinson	FZS (chimpanzee habituation attempts)	Robinson & Robinson 1998
1998-Apr-30–May-08	Anne Pusey	Duke Univ., USA (survey)	Pusey 1998
1998	D. Ommaney	FZS (habituation attempts)	Ommaney 1998
2000-Sep-02–16	Akiko Matsumoto-Oda	Primate Research Inst., Kyoto Univ., Japan (research)	Matsumoto–Oda 2000
2002–2008	Michael A. Huffman	Primate Research Inst., Kyoto Univ., Japan (research)	Huffman et al. 2008

Period	Name(s)	Institution (focus)	References
2002–2004	Liza R. Moscovice	Univ. of Wisconsin–Madison, USA (PhD research)	Hasegawa et al. 2005, Moscovice 2006 [PhD thesis], Moscovice et al. 2007, 2010
2003–2008 (intermittently)	Klára Petrželková, Lucia Bobáková, Vladimír Mazoch, Mwanahamisi Issa Mapua	Inst. of Vertebrate Biology, Academy of Sciences, Czech Republic (parasitology research)	Petrželková et al. 2006, 2008ab, 2010ab; Petrášová et al. 2010, 2011; Pomajbíková et al. 2010
2003–2008 (intermittently)	Taranjit Kaur, Jatinder Singh	Virginia Polytechnic Inst., USA (ape health monitoring)	
2012-Apr–2014-Mar	Josephine N. Msindai	Univ. College London, UK (PhD research)	Msindai et al. 2015ab, Msindai 2018 [PhD thesis], Msindai et al. 2021, Williams et al. (in prep.)
2014–2017	Felix Schürmann	Univ. of Erfurt in Gotha, Germany (archival work in Germany and Tanzania)	Schürmann 2016, 2017ab
Since 2016	Shaibu Utenga	Tanzania National Parks Authority (habituation attempts)	

Note: FZS = Frankfurt Zoological Society. Period: YYYY-MMM-DD. (Modified after Msindai et al. 2021)

However, after realising the difficulties of observing the chimpanzees, he instead collected data on the biology of four sympatric carpenter bee species (Anzenberger 1977, 1986). Once Anzenberger had moved to the University of Zurich (Switzerland), he guided a Swiss biology graduate, Guido Müller, to a 3-month study in 1994, with a focus on numbers, ranging and nesting patterns of the Rubondo apes. From 1996 to 2002, the South African couple Johan and Paula Robinson were tasked by the FZS to habituate the chimpanzees to the presence of human observers. Camps were established in the north and south of the island, and research assistants aimed to track the chimpanzees. However, habituation was not achieved and no results were published. The first detailed investigation of chimpanzee socioecology was the PhD research conducted by Lisa Moscovice (Psychology Department, University of Wisconsin-Madison, USA; supervisor: Charles Snowdon) with guidance from Michael Huffman (Primate Research Institute, Kyoto University, Japan) from Oct02 to Apr04. From 2003–04 and 2006–08, Klára Petrželková (Institute of Vertebrate Biology, Academy of Sciences, Czech Republic) ran a health monitoring study, including parasitology, together with Taranjit Kaur and Jatinder Singh (Department of Biomedical Sciences and Pathobiology, Virginia Polytechnic Institute and State University, Blacksburg, Virginia, USA), again in cooperation with Michael Huffman (Kyoto University).

The co-author of this book, Volker Sommer – a German primatologist with academic roots at Göttingen University, Germany, and extensive fieldwork experience dating back to 1981 with langur monkeys in India, lar gibbons in Thailand and baboons and chimpanzees in Nigeria – took to the idea of research at Rubondo in the early 1990s, after having met Gustl Anzenberger. Sommer contacted Grzimek's successor as FZS president, Richard Faust, but preliminary plans where scuppered when Faust passed away in 2001. Meanwhile, Sommer had moved to the Department of Anthropology, University College London (UCL) in the United Kingdom. Here he met this book's main author, Josephine Nadezda Msindai, and they developed plans for her PhD research on Rubondo with a focus on chimpanzee ranging, nesting and population history. Msindai has roots in Tanzania and Russia – a family connection that came about because of the close alliance of the socialist East African nation with what was then the Soviet Union. She had learnt about the existence of the Rubondo apes during earlier fieldwork with primates in Africa – Barbary macaques in Morocco, guenons and vervet monkeys in Kenya, chimpanzees in Uganda, highland mangabeys in Tanzania. As a result, research on Rubondo took place over 22 months, from Apr–Aug12, and Nov12–Mar14. Co-author VS joined the fieldwork efforts in May–Jun12, Sep–Oct12 and Jul–Aug13.

The research team was composed of seasoned trackers and volunteer researchers (Figure 2.5). Particularly important was the expertise of James Leonard, Joseph Mgwesa and Simon Nkuzi who hailed from nearby Mwanza city. Given their existing connections to TANAPA staff, they were initially recruited in the late 1990s by FZS to habituate the chimpanzees for tourism

Figure 2.5 Team members of the author's research team (Jul13). FA = local field assistant, CC = camp cook, SR = senior researcher, DR = doctoral researcher, VO = volunteer. From left to right top: CC Damali Obiedy, FA Simon Nkuzi, DR Josephine Nadezda Msindai, FA Musa Mwenge, FA Samson Obiedy, VO David Dietz, FA James Leonard; bottom: SR Volker Sommer, FA Onesmo Mazazele, FA Joseph Mgwesa, SR Paco Bertolani. (Photo: VS)

and research, under the guidance of Paula and Johan Robinson. They went on to assist research on the spotted-neck otter and vervet monkeys. The base camp on Rubondo was a three-room house, originally constructed by Ulrich Kade, equipped with cold running water, electricity (powered by a generator from 19:00 to 23:00 hrs) and a gas stove. A boat, provided by TANAPA, enabled trips to the mainland. This allowed the team to buy consumables such as food and drinks at Mganza village market. Petrol was bought in Geita town, again aided by the park management. To move between the camp and the study areas, two Yamaha 100 cc and later four Honda 110 cc motorbikes were used.

Much of the funding on the ground was provided by Asilia, a tourism company operating 19 lodges in Tanzania and Kenya, via the NGO Honeyguide Foundation. The cooperation was based on the mutual interest to habituate Rubondo's chimpanzees to humans for research and tourism. Alas, the study was not long enough to achieve this, and continued efforts since 2016 by Asilia and TANAPA have likewise made slow progress to date.

Founders of a Population: Numbers, Sex, Origins

While there is no dispute *that* chimpanzees have been released on Rubondo, conflicting information has been provided about the origins, numbers, age and sex of the released animals, as well as their subsequent fate. In the following, we trace various sources and causes of confusion, reconstruct the provenance of released chimpanzees and point out where information is lacking. Much of this is based on the tabulated overview about the founders of the Rubondo population provided above (c.f. Table 2.1). When summarizing the emerging picture, we consolidate some information that we have detailed in previous sections.

As already mentioned, the project centres around 17 chimpanzees (7 males, 10 females) previously held captive in Europe and likely born between 1955 and 1962, of which 16 were released (7 males, 9 females). The chimpanzees were transported to Africa in four successive waves, with male–female ratios of 3:8, 1:0, 1:0 and 2:2.

- The first cohort left Antwerp on 21May66 on board a freighter. En route, 1 female chimpanzee, Joséphine, died (no. 3; cf. Table 2.1). The remaining, 3 males and 7 females, were offloaded in Dar es Salaam and released on Rubondo on 23Jun66.
- The second release was of a single male on 19Oct66.
- The third release was of a single male sometime in Jan68.
- The final and fourth release of 2 males and 2 females occurred on 30Jun69.

Previous publications agree on the number and composition of the three last waves. However, statements about the first wave are often incorrect. This is at least partly related to the fact that 11 animals (3:8) were shipped out from Europe, but – because 1 female died during the journey – only 10 animals were released (3:7).

Interestingly, Grzimek himself, in one and the same publications, goes back and forth between 10 and 11 animals released, e.g. during the TV documentary (Grzimek 1966a), in a German book account (1969: 14, 37), its English translation (1970: 13, 36) and a later redacted version (Grzimek 1988: 482, 485). It appears unlikely that Grzimek made an honest mistake when explicitly stating that he "dispatched ten large chimpanzees from Antwerp" (Grzimek 1971: 13; ditto 1969, 1988). Instead, one is tempted to assume that he tried to keep the unpleasant fact of the death of one animal out of the public domain. Still, in a message to members of the FZS, he justifies the transport of zoo animals to Africa by arguing that catching wild apes would have necessarily entailed the death of some of those pursued, "while of the chimpanzees brought to Africa from zoological gardens, only one animal died during the initially extremely stormy sea voyage" (cit. in Sewig 2009: 317f). There is no mention that Joséphine died due to a heat stroke.

Be this as it may, subsequent researchers and publications have continued to erroneously report that a total of 17 chimpanzees were released – instead of just 16, e.g. Borner (1980, 1985, 1988), Müller & Anzenberger (1995: 12), Matsumoto-Oda (2000), Moscovice et al. (2007), Huffman et al. (2008), Petrášová et al. (2010). Our own previous work (Msindai et al. 2015) also uses the wrong information, albeit we documented the correct figure in a follow-up publication (Msindai et al. 2021).

Other publications, in addition to erroneously stating that 11 animals were set free in the first wave, at times also misrepresent the sex ratio. Instead of stating it as 3:8 (which would include the deceased female), it is given as 4:7 (e.g. Borner 1985, Müller & Anzenberger 1995, Huffman et al. 2008). Moreover, while correctly reporting that 10 animals were released during the first wave, their sex ratio is wrongly stated as 4:6 instead of 3:7 by Schürmann (2017b: 291) because an unnamed male was in fact a female (Kathrin, no. 9; cf. Table 2.1). Finally, Müller and Anzenberger (1995: 12–14) tabulate 12 animals being shipped to Africa during the first wave (4:8, including the female that died in transit). However, cross-checking could not confirm the existence of an unnamed male (no. 9, likely an erroneous duplication of an entry; Gustl Anzenberger, pers. comm. to VS), which is therefore excluded from our tabulation.

What do we know about the origin of the founders of the Rubondo population? Common chimpanzees (*Pan troglodytes*) are divided into subspecies from West Africa (*P. t. verus*), Nigeria–Cameroon (*P. t. ellioti*; previously *vellerosus*), Central Africa (*P. t. troglodytes*) and East Africa (*P. t. schweinfurthii*), while their co-geners, the bonobos (*Pan paniscus*) are only found south of the Congo River (Chapter 3, Chapter 6). Previous publications made various conjectures about the geographic provenance of the Rubondo founders – assumptions that would translate into their origin from a particular subspecies – which were then seemingly carried over from one source to the next. Thus, several sources explicitly state that all released apes were born in the wild in Africa (Borner 1985: 152, Pusey 1998: 2, Matsumoto-Oda 2000: 16, Moscovice et al. 2007: 3, Huffman et al. 2008: 222, Msindai et al. 2015). A more specific origin from West Africa is also made explicit (Pusey 1998: 2, Matsumoto-Oda 2000: 16, Huffman et al. 2008: 222, Petrášová et al. 2010: 923) while some sources mention the individual countries Guinea, Ivory Coast and Sierra Leone (Matsumoto-Oda 2000: 16, Huffman et al. 2008: 222).

Information from the archives about the provenance of the 17 chimpanzees sent from Europe to Africa, while indeed sketchy, nevertheless calls for a revision of previous statements (cf. Table 2.1). There is no origin record at all for 4 of the apes. For 9 of the remaining, 'Africa' resp. 'probably Africa' is indicated, while for only 4 individuals, a specific country is noted (1x Sierra Leone, 2x 'probably' Guinea, 1x 'probably' Ivory Coast). Importantly, 1 individual of 'known' origin – said female Joséphine ('probably' from Guinea) – died in transit. This leaves just 3/16 (19%) of the founders with a designated regional origin, i.e. an unnamed female from Sierra Leone, female Kathrin/Caroline 'probably' from Ivory Coast, and female Schilla, 'probably' from Guinea.

However, even those designations only indicate the country from where they were shipped out and not necessarily, where they were born, given that "assumptions about the geographical (or 'subspecies') origin of captive populations based on the site of purchase can be very misleading" (Gagneux et al. 2001). This caveat refers to the 35 founders of the Rijswijk chimpanzee colony that had been purchased from a dealer in Sierra Leone, and it was therefore presumed that they originated from upper Guinea. However, mtDNA genotypes revealed that two were, in fact, of Central African origin (ibid.).

Given the complete lack of origin information for 4 founders, it even seems theoretically feasible that they were born in captivity instead of in the wild. However, we can virtually exclude this possibility, because according to the studbook (Carlsen & de Jongh 2014), during any year before the 1970s at the most only a handful of chimpanzees were bred in European zoos. In addition, names or other details of the 4 founders born 1959–60 for whom we lack origin records (nos. 14, 15, 16, 17, cf. Table 2.1) do not match the only 2 animals listed as captive born during that time (ibid.).

Hence, archival records are limited in that they are suggestive, but not definitive. We therefore utilized biological samples collected during our fieldwork to employ genetic analyses and explore subspecies identities via matrilines of Rubondo chimpanzees. This work was conducted at the German Primate Center, Göttingen, under the direction of Christian Roos, with assistance by JNM (details in Msindai et al. 2021). DNA was extracted in 2015, 12–36 months after collection, from a total of 201 samples (faeces n=196, hair n=4, food-wadges n=1). For this, mitochondrial DNA was separated with the QIAquick Gel Extraction Kit (Qiagen), as outlined in Kalbitzer et al. (2016). Exploiting published primers, we amplified and sequenced a 460–500 bp-long fragment of the mitochondrial control region of the hypervariable region I (HVI). Once sequences were obtained, they were aligned in SeaView 4.5.4. Using an existing set of sequences of known chimpanzee subspecies, we assigned an identity to each sample. Molecular sex determination of samples was conducted using a universal PCR-based sexing system in which fragments of the X-chromosomal DDX3 gene and of the Y-chromosomal DDY3 gene are simultaneously amplified (Ferreira da Silva et al. 2018).

In total, HVI sequences (491–492 bp) were successfully generated for 196 samples, with 116 derived from females, 79 from males and 1 sample of undetermined sex. Among these, we found 4 haplotypes. Using BLAST search, one of them was identified as *P. t. troglodytes* (haplotype troglodytesA) and three as *P. t. verus* (haplotypes verusA, verusB and verusC) (Figure 2.6).

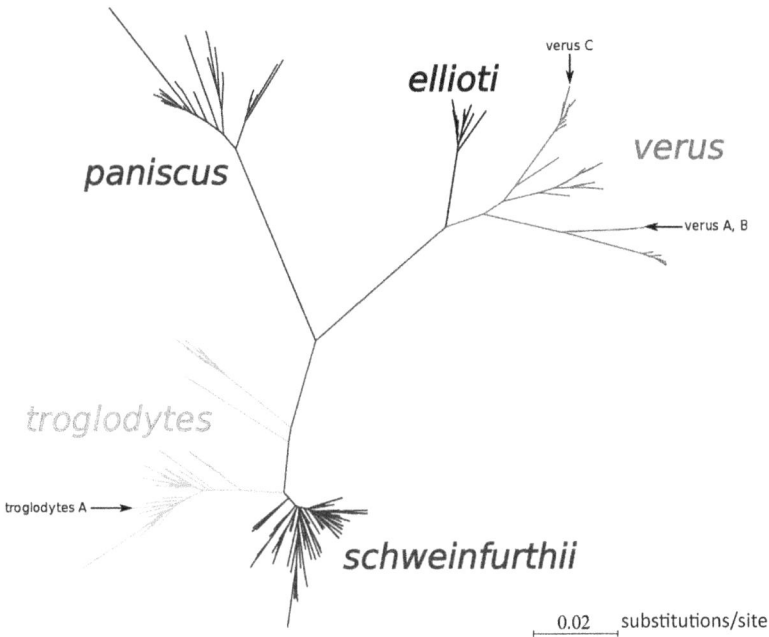

Figure 2.6 Phylogenetic relationships among chimpanzees and bonobos, including haplotypes of the Rubondo chimpanzees (modified after Msindai et al. 2015). Common chimpanzees (*Pan troglodytes*) are divided into subspecies from West Africa (*verus*), Nigeria–Cameroon (*ellioti*; previously *vellerosus*), Central Africa (*troglodytes*) and East Africa (*schweinfurthii*), while their co-geners, the bonobos (*Pan paniscus*) are only found south of the Congo River (see Chapter 4).

Haplotype troglodytesA, found in 66 individuals (23 males, 43 females), differs from its closest related sequence in Genbank (*P. t. troglodytes*: JN191203.1) in four positions (99.2% identity) and in at least 13 positions (97.4% identity) from sequences of other *P. t. ssp.* Haplotype verusA, found in 2 individuals (both females), is 100% identical to the *P. t. verus* sequence JN191232.1 and differs in at least 26 positions (94.7% identity) from other *P. t. ssp.*, while the verusB haplotype, found in 80 individuals (41 males, 38 females, 1 undetermined sex), differs from haplotype verusA and JN191232.1 in one position (99.8% identity) and from other *P. t. ssp.* in at least 27 positions (94.5% identity). The verusC haplotype, found in 48 individuals (15 males, 33 females), differs from its closest related sequence in Genbank (*P. t. verus*: FJ642360.1) in three positions (99.4% identity) and from other *P. t. ssp.* in at least 29 positions (94.1% identity).

In summary, 33.7% of samples (23 males, 43 females) derived from individuals with a *P. t. troglodytes* haplotype and 66.3% of samples (56 males, 73 females, 1 undetermined sex) from individuals with a *P. t. verus* haplotype.

Two decades before our own analyses, 6 chimpanzee hairs collected from night nests on Rubondo (1994 fieldwork, Müller & Anzenberger 1995) had already been genotyped via mitochondrial DNA (1996 lab work, Pascal Gagneux, University of California at San Diego), yielding data for 2 individuals. The results had remained unpublished, but because all hairs were from animals of the *P. t. verus* subspecies (Gustl Anzenberger, pers. comm. to VS), they are in accordance with our own findings.

The combined archival and genetic evidence therefore leaves no doubt that at least some, and probably the majority of the Rubondo founders, belonged to *P. t. verus*, the subspecies which inhabits Western Africa, mainly Côte d'Ivoire, Guinea, Liberia, Mali and Sierra Leone, but with populations in surrounding countries. Our genetic analyses also indicate that populations of *P. t. troglodytes* contributed to the Rubondo genepool. This subspecies inhabits Central African countries, with main populations occurring in Cameroon, Gabon and Republic of Congo. There was no evidence for provenance from the Nigeria–Cameroon subspecies *P. t. ellioti* – nor from East African *P. t. schweinfurthii*. (Note: We ignore the unlikely possibility that these matrilines were originally present but died out in the meantime.)

In parallel to the Rubondo situation, the overall historical records for 3,906 chimpanzees held in European institutions till 2014 for which subspecies were known (n = 906) display a clear preponderance of *P. t. verus* (85.1%) over *P. t. troglodytes* (13.2%), with near absence of *P. t. schweinfurthii* (1.4%) and *P. t. ellioti* (0.2%) (Carlsen & de Jongh 2014: 36). The extant 2016 captive population comprises 25.5% animals of unknown provenance and 34.2% hybrids, while 40.4% of the apes are genotyped to a distinct subspecies level. Similarly, these 310 animals are 82.3% *P. t. verus*, 13.9% *P. t. troglodytes*, 3.9% *P. t. schweinfurthii* and 0.0% *P. t. ellioti* (Frands Carlsen & Tom de Jongh, pers. comm. to JNM; see also Hvilsom et al. 2013, Fisken et al. 2018, Frandsen et al. 2020).

This pinpoints to chimpanzee habitat in the immediate mainland interior along the Atlantic coast as the preferred area of capture to obtain apes for the European captive market – a dynamic also reflected in the Rubondo population. The lack of *P. t. ellioti* might be explained by the comparatively smaller alignment of the subspecies habitat with the coastline in Nigeria and Cameroon (Oates et al. 2009). The absence of East African founders amongst the Rubondo apes is likely related to the fact that – in contrast to the situation along the Atlantic – the coast of the Indian Ocean in East Africa is quite distant from the contiguous habitat of *P. t. schweinfurthii*, which consequently requires long transport of captured apes to the next harbour. In fact, Grzimek alludes to the preferred trade route when mentioning that "most chimpanzees are captured" at the "West coast" (Grzimek 1971: 16).

Similarly, the studbook ascribes an East African origin to only 12 captives captured between ca. 1900 and 1974.

Current guidelines of the International Union for the Conservation of Nature (IUCN) recommend that different subspecies should not be released jointly to preserve genetic diversity (Beck et al. 2007; Chapter 6). However, it should be kept in mind that genetic testing became only recently possible and affordable, and that taxonomy is notoriously subject to change. When Grzimek released the first cohort in Rubondo in 1966, the subspecies division of *Pan troglodytes* was only just taking hold in academic circles (Napier & Napier 1967). Still, while Grzimek was acquiring chimpanzees, George Pournelle of San Diego Zoological Garden, USA, suggested to him that he should only bring East African chimpanzees and bonobos to Rubondo, to keep hybridization low and also because these taxa were particularly rare (letter from Jul65, FZS archives, Afrika, 00001976, Rubondo 1, Keller Regal 01019; Felix Schürmann, pers. comm. to VS). Ergo, Grzimek must have been aware of potential negative perceptions to release apes from the other side of Africa onto Rubondo, although he obviously ignored it – if only because neither East African chimpanzees nor bonobos were held captive in Europe at the time.

A 'Homecoming'? Or 'Jurassic Park'?

Grzimek provides a noteworthy résumé of the whole operation: "For the first time, anthropoid apes from Europe had travelled back to their original home in Africa" (Grzimek 1971: 35). While the generic destination 'Africa' is correct, the specific location of Rubondo is certainly not the 'original home' of the released animals. Dunnett, in 1969, also talked about "chimpanzees sent back to their natural home" (cit. in Schürmann 2017b: 287). These narratives of a homecoming are quite different from when Peter Achard, back in 1961, likened the introduction of new species to the creation of an "artificial game reserve". Nowadays, given the popularity of the media franchise surrounding the novel by Michael Crichton (1990), one would be tempted to draw comparisons to the fictional isle of 'Jurassic Park', which was likewise populated with an artificial array of animals.

Still, whether romanticizing or rationalising the operation, such labels became academic after the primates had been released – because the apes were here to stay. Grzimek (1971: 35) had concluded the recount of his "experiment" with an emphatic wish – which, at least for the surviving releasees, should become reality: "Good luck to the chimpanzees of Rubondo!"

3 Rubondo Island: Weather, Forests, Wildlife, Humans

Figure 3.1 Aerial view of Rubondo Island National Park's coastal area and forest cover. (Photo: VS)

Understanding animal behaviour requires us to understand the environmental conditions that exert selective pressure on the study population. Hence, to contextualise our findings about the Rubondo chimpanzees (Chapter 4, Chapter 5), we provide an overview on habitat features they have to negotiate, particularly its island character, climate, vegetation cover, sympatric fauna and human impact. Abiotic factors such as terrain and topography, precipitation and temperature as well as biotic factors such as plant

DOI: 10.4324/9780367822781-3

phenology and coexisting wildlife constitute a baseline against which patterns of ranging, foraging and night nests construction can be examined. Such baselines also inform comparisons with endemic chimpanzee populations elsewhere. Still, within this framework, Rubondo in its well-protected offshore location (Figure 3.1), stands out as one of the rare habitats almost unaffected by the destructive anthropogenic activities that are increasingly shaping the globe.

Lake Victoria: Forming, Shrinking, Expanding

Rubondo Island National Park (Figure 3.2), gazetted in 1977 (Chapter 1), is located in the south-western corner of Lake Victoria, Tanzania (latitude 2 18' 10.3", longitude 31 51' 26.9", altitude 1,100 m). The main island encompasses 237 km^2, and is surrounded by another 11 islets, none much larger than 2 km^2. The surrounding waters are also gazetted, which brings the total protected area to 457 km^2.

The park consists of a partially submerged rift of four volcanically formed hills, linked by three flatter isthmuses. The highest point, 400 m above the level of the lake, is Msasa Hills in the far south at 1,486 m. There are no rivers, but several creeks are fed by rainwater stored in crevices of sandstone rock. North to south, the island measures 28 km, while the width varies from 2.5–10.6 km. About 80% of the main island is covered in forest.

The surrounding Lake Victoria straddles the equator from 0.5° N to 3.0° S, sitting in the depression between the eastern and western branches of the East African Rift Valley System (EARS) (Chorowicz 2005). The lake's area is divided among Kenya (6%), Uganda (45%) and Tanzania (49%). Known in African-language names as Ukerewe or Nnalubaale, the water body's English name was chosen in 1858 by British explorer John Hanning Speke to honour the then Queen Victoria of England (Jeal 2012) – a colonial approach of naming a major entity of a global landscape to which there hasn't been much opposition since. No matter what, Lake Victoria (for the following, see Tryon et al. 2016) is the biggest African lake as well as the biggest tropical lake in the world. The Kagera River constitutes its largest tributary (10% of total inflow) and drains the highlands of Rwanda and Burundi to the west. Outflow is to the north at Jinja, where Lake Victoria serves as a primary source of the Nile.

With a surface area of 68,800 km^2, it is roughly the same size as Ireland. As a defining geographic feature of Equatorial Africa, it is so vast that it generates its own weather system. Average rainfall across the lake is equal to average evaporation levels. The mean annual precipitation varies at different points around and on the Lake itself from 1,200–1,600 mm, with higher values recorded in the north and lower amounts in the south (Figure 3.3). Because local precipitation comes primarily from the lake itself, the surface expands and contracts in response to changes in rainfall.

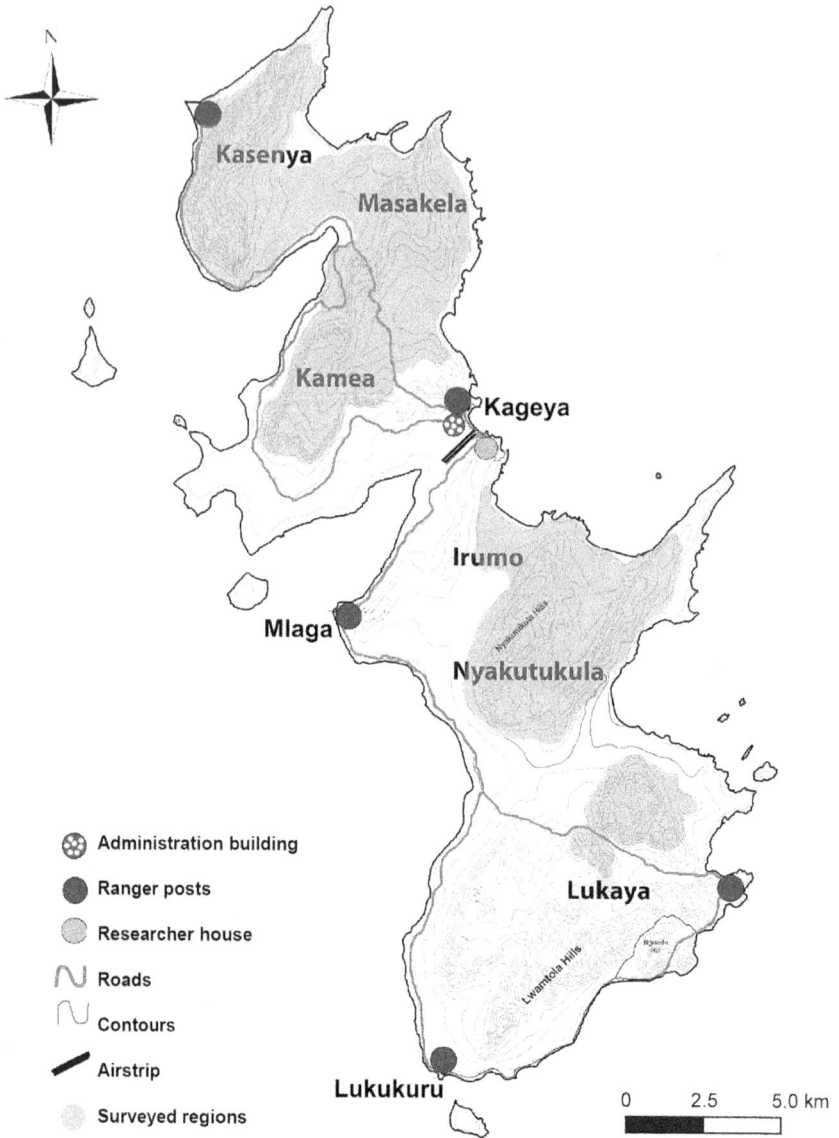

Figure 3.2 Rubondo Island National Park and important locales of its main island. Note: Other sources may use slightly different spellings, such as "Kageye". (Map: JNM)

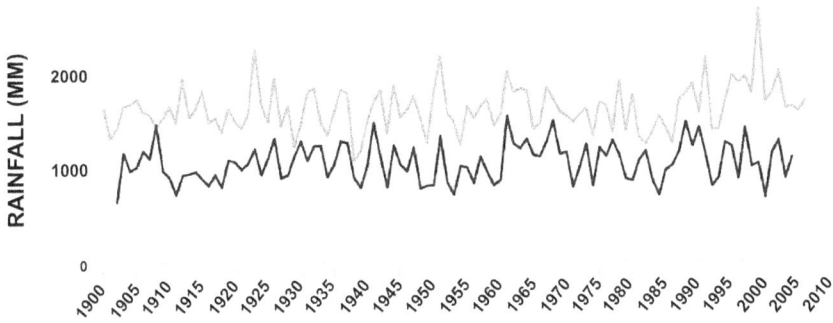

Figure 3.3 Annual rainfall 1900–2006 at meteorological stations on Lake Victoria. Top line Entebbe (Uganda), bottom line Mwanza (Tanzania). (Modified from atlas.nilebasin.org)

Historical water level fluctuations of ca. 4 m have been documented in the 19th century, while the high precipitation of 1962, following the end of colonial rule in late 1961, is charmingly remembered as 'independence rains' by Tanzanian people (Kiwango et al. 2005). Although wide, the lake is shallow with a maximum depth of only about 83 m. As a result of this geometry, large changes in lake surface area can happen with modest changes in depth.

Geologically, the lake's basin is relatively young and formed only 400,000 years ago in the mid-Pleistocene. In its early stages, pre-lake rivers fed into swampy lakes, which then gradually interconnected, losing their outlet to the west. Paleolimnological evidence points to major fluctuations in water level, caused by reduced rainfall, including complete or near-complete desiccations about 16,000 and 14,000 years ago, intermittently reducing the surface area to less than a tenth of its current size (Stager et al. 2011).

In terms of fossil fauna (Faith 2014), extinct taxa account for >50% of the large mammals, but many extant species are also identified, such as baboon, lion, bushpig, zebra and a type of wildebeest. However, fossils of chimpanzees have not been found so far, which coincides with their current absence from the lake's shores. As the lake sits at the junction between central African forests to the west and savanna habitats to the east, it forms an important boundary for populations of large mammals. Because of this, classic forest species such as red river hog, okapi and chimpanzee are found to the west, while savanna species such as wildebeest, zebra and warthog thrive to the east. The replacement of the lake with grassland during the desiccation periods would have removed the dispersal barrier, expanding the range of at least the grazing herbivores (ibid.).

When the lake nearly dried up 16,000 years ago, some fish survived, including cichlids, a group of small perch-like fish which originated some 130,000 years ago from intermixing of types from the Congo River and the Upper Nile River watershed. The lake's desiccation separated the cichlid populations, which led to independent evolutionary changes until the lake filled again 12,000 years ago. This spurred a remarkable radiation into about 500 species, with some feeding on detritus while others are algae scrapers, snail crushers, ambushing or fast-swimming predators (Goldschmidt 1996).

About 10,000 years ago, at the onset of the Holocene, the lake overflowed via the Victoria Nile, draining northwards (Beuning et al. 2002). For this time, pollen records are dominated by trees in the fig family (Moraceae), signifying the growth of forests and a return to humid conditions. Today's coastal vegetation is predominantly bushland, thicket, savanna-woodland and forest, comprising a mosaic of often endemic taxa (Tryon et al. 2016).

Island Climate: Rainfall, Temperature, Humidity

Weather records for 1977–96 were obtained from the FZS, courtesy of Markus Borner, and for 1997–2013 from TANAPA, courtesy of park ecologist Wickson Kibasa. At Kageya park headquarters, on the eastern side of the island, an electronic weather station was used. At other ranger posts, manual records were obtained via standardised rain gauges and a digital thermometer. While care has been taken, for logistical reasons – such as broken equipment or absence of staff – some data are missing for certain periods of time.

In any case, long-term values, measured at Kageya (Table 3.1), reveal a rather stable weather pattern across the year, as indicated by the small ranges of monthly fluctuations (mean maximum temperature 28°C, range 26–29°C; mean minimum temperature 17°C, range 15–19°C; humidity 57%, range 47–63%). The hottest day saw 41°C, the coolest 11°C. On average, Rubondo experiences 30% rainy days per month, which corresponds to 110 rainy days per annum. Maximum rain recorded on a single day was 106 mm, and the sum across the monthly means comes to 1,289 mm.

A breakdown of the data for an 11-year period (2004–14; Figure 3.4) reflects the characteristic seasonality in north-western Tanzania, with two rainy seasons, typically Mar–May and Oct–Dec, in which rainfall exceeds 100 mm per month. There is a short dry season, Jan–Feb, and a longer dry season, Jun–Sep, with rainfall below 50 mm per month.

Average yearly rainfall, as measured at Kageya, reveals considerable interannual variation (Figure 3.5). As a result, there can be spells of

Table 3.1 Main climate variables on Rubondo (2004–14)

Variable (2004–14)	Mean	Jan	Feb	Mar	Apr	May	Jun	Jul	Aug	Sep	Oct	Nov	Dec
Max. temperature (°C) (a)													
Average across years	28	29	29	28	27	29	28	28	29	26	26	26	27
Min. monthly mean	24	27	25	23	23	25	23	21	25	23	23	22	24
Max. monthly mean	30	32	31	30	29	31	31	32	33	29	28	29	30
Highest daily value	34	35	34	33	33	32	41	33	33	36	31	33	32
Lowest daily value	23	22	27	26	25	28	26	21	19	23	19	21	19
Min. temperature (°C) (a)													
Average across years	17	17	17	19	18	19	17	16	15	17	17	17	17
Min. monthly mean	15	13	14	17	17	17	14	13	11	16	16	17	16
Max. monthly mean	20	20	23	23	23	24	23	18	18	18	18	18	19
Highest daily value	20	22	21	19	20	21	20	21	22	19	20	19	21
Lowest daily value	13	13	13	14	15	14	13	11	11	14	13	13	14
Average temperature (°C) (a)													
Mean of average min and max	22	23	24	23	23	24	23	20	23	22	22	22	23
Min. monthly mean	17	17	18	18	19	19	17	11	16	17	17	17	17
Max. monthly mean	28	29	29	28	27	29	28	29	29	27	27	26	28
Humidity (%)													
Monthly mean	57	58	60	63	62	52	54	44	47	60	58	66	64
Rain (mm) (b)													
Mean	107	98	93	197	163	124	24	3	29	92	125	165	176
Highest daily value	106	71	55	105	81	106	34	11	55	73	74	54	68
Lowest daily value	0	0	0	0	0	0	0	0	0	0	0	0	0
Min. total mean	21	29	0	66	17	0	0	0	0	16	7	35	84
Max. total mean	221	202	198	431	280	282	110	11	175	175	228	326	344
Mean rainy days/month (%)	30	28	32	37	41	27	8	5	11	33	42	48	43

(a) No data for: 2004–07; Apr09–Apr10; Jun10–Jan12
(b) No data for Feb–Dec07, Sep–Dec10

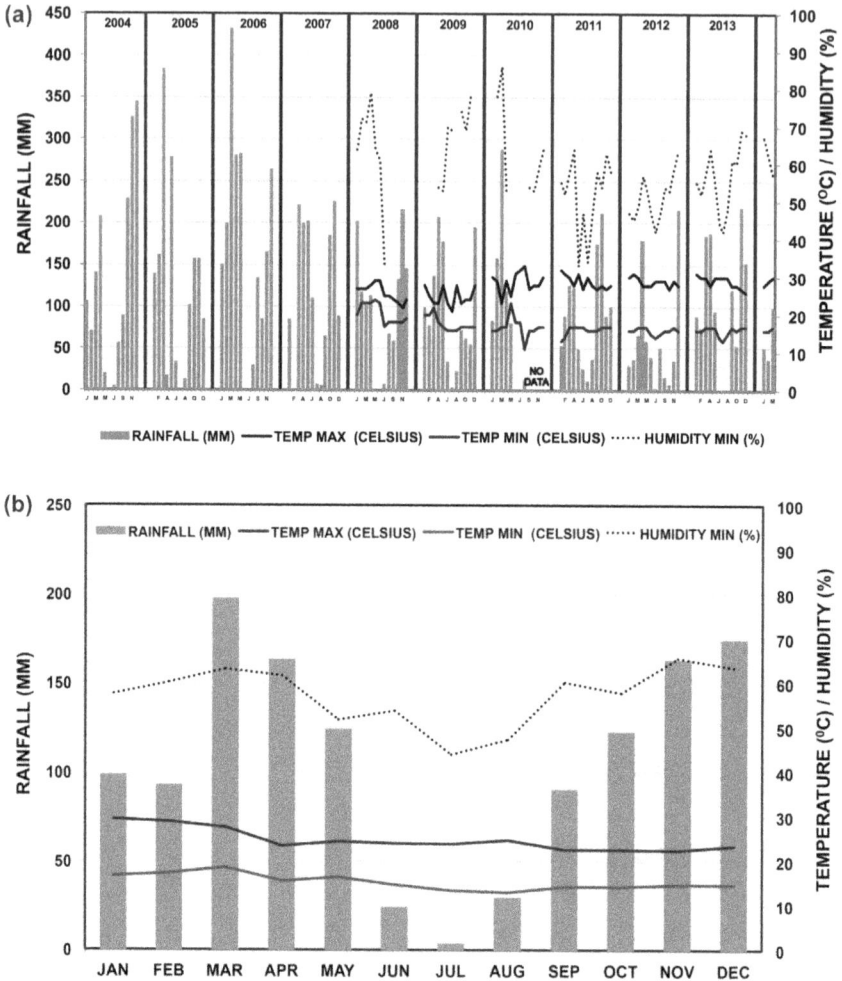

Figure 3.4 Weather on Rubondo, 2004–14. (a) Yearly values. (b) Averaged across 11 years.

relatively wet years (mean 1,709 mm, 2004–06) and dry years (999 mm, 2011–13), which translates into a 40% reduction. However, according to a 2012 assessment of John Msindai, geologist at the Open University of Dar es Salaam, the island is partly composed of laterite and quartz-sandstone rock. Sandstone readily absorbs rain because of its porous nature, and can store large quantities, making it a valuable aquifer for plants in drier times.

Within a sample of 14 chimpanzee study sites, median annual rainfall is 1,678 mm (compiled from Hunt & McGrew 2002: 40, Pruetz & Bertolani

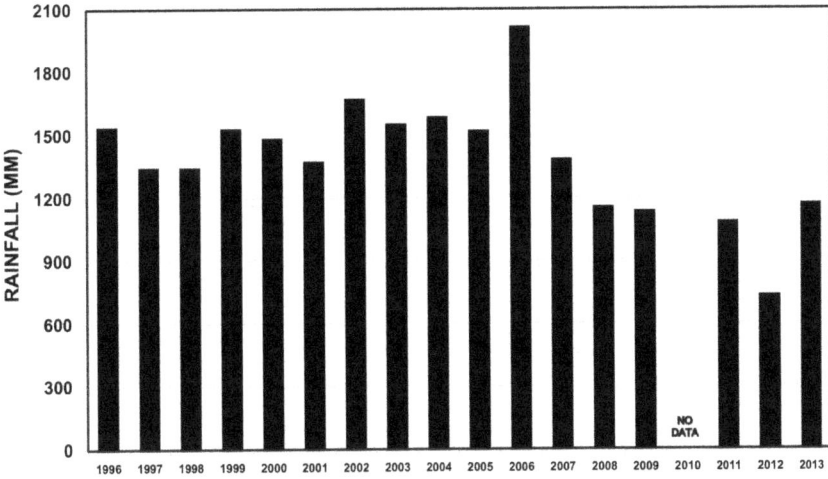

Figure 3.5 Total annual rainfall over the 18-yr period 1996–2013.

2009, Jesus 2020). According to this comparison, Rubondo is the third most dry site:

- Assirik, Senegal 954 mm;
- Issa, Tanzania 955 mm;
- Rubondo, Tanzania 1,289 mm;
- Semliki, Uganda 1,379 mm;
- Lope, Gabon 1,531 mm;
- Kibale, Uganda 1,671 mm;
- Budongo, Uganda 1,684 mm;
- Gombe, Tanzania 1,775 mm;
- Kahuzi-Biega, DR Congo 1,800 mm;
- Taï, Ivory Coast 1,829 mm;
- Mahale, Tanzania 1,836 mm;
- Gashaka, Nigeria 1,968 mm;
- Bossou, Guinea 2,230 mm.

However, while Rubondo is relatively dry, there is only a mild pattern of seasonality compared to other study sites. E.g. in Gashaka, Nigeria, a dry spell of 5-months is followed by heavy rains when monthly maxima can reach more than 500 mm, making up 97% of all precipitation (Jesus 2020).

The reassuring take-home message for chimpanzees as well as the island's other fauna and flora (if they could read it...) would be that there is no need to fear extreme weather fluctuations, neither intra- nor inter-annually. Hence, while the biblical proverb maintains 'To everything there is a season', Rubondo's rule is more like: 'Not everything has a season'.

Guido Müller – on Rubondo in 1994

Figure B3.1 Guido Müller – on yet another rainy day on Rubondo. (Photo: Guido Müller)

Looking back, it was all fantastic! Markus Borner dropped me off in his famous zebra-striped plane, said "see you again in three months" and flew away. Standing there with my backpack, I felt like Robinson Crusoe. This was before internet and mobile phones, so it felt very remote indeed. Initially I felt lonely, lacking human contact, and had little in the way of food to eat. But things improved with time. I slept in my tent, until the rangers allowed me into the Borner house, where I found a shelf full of old tinned food. That was a welcome relief because I had lost 10 kg of body weight by then; once I had access to the food store I ate more regularly. The rangers didn't speak English, and I only mastered a few words in KiSwahili. So, when I was told one evening the trip to the market on the mainland would start at 'saa kumi na moja asubuhi' (at eleven in the morning), I thought this a little late. Only the next day the rangers woke me up before sunrise. I then realised that they count the hours from 6 in the evening – thus 11 is 5 in the morning. My ambition was to estimate the number of chimpanzees, determine their ranging pattern, map nesting sites and create a herbarium with food plants. But, not least, because it was the rainy season, I often struggled to find the chimpanzees. I was a little naïve, thinking I could just go there and be like Jane Goodall – live amongst wild animals. I quickly realised that the reality on the ground is very different; but hey, Rubondo certainly taught me great lessons for my future professional life.

Forests: A Pantry of Fruit

Animal behaviour and reproduction is constrained by habitat features such as the texture of geology and climate, but also vegetation cover, which can provide shelter and food. Plants themselves are strongly influenced by the environment, and the discipline of phenology explores these effects with respect to their life cycles (Osborne et al. 2000). The extent and structure of plant cover is impacted by annual resp. supra- or sub-annual cycles in rainfall (Adamescu et al. 2018). The timing of rain and sunlight determines the succession of phenophases, i.e. leafing, flowering and fruiting which govern the reproductive success of plants. In turn, animals have to respond to the cyclical events in plants in terms of the quantity, quality and distribution of items that provide food or protection (van Schaik et al. 1993).

As for primates, numerous studies confirm that the patterning of behaviour and social organisation largely reflect the availability of food resources, especially fruit (Chapman et al. 2017). However, compared to forests in Asia and the Neotropics, the phenology of African forests is not well understood (Adamescu et al. 2018). The main research challenges are variations in phenophases between and within habitats, and a general paucity of long-term datasets.

Within this framework, we provide background information about climatic patterns and corresponding changes to the fruit production of trees and associated vines on Rubondo. This material prepares the ground for a broader goal of our research, i.e. linking habitat ecology to chimpanzee sociality, nest building and foraging (Chapter 4, Chapter 5).

The basic vegetation cover of Rubondo Island is mixed evergreen lowland forest as well as deciduous forest on the slopes, interspersed with patches of open grassland, swamp and woodland. The eastern lakeshore is characterised by rocky areas and sandy beaches, while the western shore supports extensive papyrus swamps lined with date palms.

Methodology: Transects and Phenological Monitoring

The composition and phenology of trees and lianas (aka woody climbers or vines) in important areas favoured by chimpanzees (Figure 3.6, Chapter 4) was measured by two different sets of researchers, firstly in 2001–03 by Liza Moscovice (Moscovice 2006: 69–113, Moscovice et al. 2010), and secondly in 2012–14 by our own team. The methodology was different in these surveys.

The 2001–03 study concentrated on three areas of predominantly semi-deciduous forest habitat where chimpanzee evidence was concentrated (Kasenya, far northern part of the island; Masakela and Kamea, northern part; Nyakutukula, central part; cf. Figure 3.2). Here, 9 transects were established of 10 m width and 300–500 m length. Tree density and composition was measured in 86 quadrats of 25x10 m, placed at 25 m intervals along transects. Within each quadrat, all trees ≥ 10 cm DBH (diameter at

Figure 3.6 Interior forest on Rubondo. (a) Section with trees of smaller diameter. (b) Trees with tangling lianas. (c) Open growth with buttressed trees. (Photos: JNM)

breast height, 1.3 m above ground) were tagged, which yielded a sample of 1,096 trees across a survey area of 2.15 ha. A subset of 6 quadrats per transect (n = 54 total) was selected for phenological monitoring and also to identify and monitor lianas. Identification to species level, wherever possible, was undertaken by Frank Mbago (herbarium of Botany Department, University of Dar es Salaam).

For our 2012–14 study, we created two perpendicular straight-line forest transects of 4 km each in Oct12 in the island's north. The chosen transect locations – east-west in Kamea, north-south in Masakela – incorporated elevation gradients from hilltop to valley. The aim was to include roughly 1,000 trees with a minimum DBH of 10 cm. Working with a transect width of 2 m, this approach resulted in a final sample of 1,094 trees with 640 associated lianas. Scientific names were assigned in situ to the trees in Feb14 by Simon Nkuzi, a trainee of botanist Frank Mbago. The trees were labelled with standard markers imprinted with successive numbers. Labelling lianas proved impractical because, when there were several stems attached to a tree, it was difficult to ascertain if a climbing plant tagged at eye level was indeed the same as that seen flowering or fruiting in the canopy. Hence, instead of tagging individual lianas at eye level, records were collected when they produced green leaves, fruits or flowers in a tree canopy. Field assistants James Leonard and Joseph Mgwesa were trained in phenological monitoring from Nov–Dec12 until their independent scoring had reached a low variation (10%). From Jan–Dec13, data were collected twice per month, on days 01–04 and 15–18, recording presence or absence of fruit, flowers and leaves. Fruit quality was assessed by noting down the size and colour, though these variables do not necessarily indicate ripeness. Fruit remains in the canopy or on the forest floor were also noted, along with likely identities of arboreal and terrestrial fruit and seed consumers. Tree height, and height of the lowest two branches was recorded via a clinometer (resolution: 0.1 m <100 m or 1 m >100 m). DBH was measured at 1.2 m above the forest floor. A 'fruit index' for each month was calculated by summing up the DBH of all fruit-bearing plants on the transect. DBH is a good indicator of both tree height and crown diameter. These in turn are good predictors for the size of fruit crops (Leighton & Leighton 1982, Chapman et al. 1992).

We report mostly on our own 2012–14 findings, with some cross-referencing to the 2001–03 study.

Forest Architecture

The 2012–14 study established that almost two-thirds of all trees (63%) had a trunk diameter of 10–20 cm (mean = 16 cm, median = 12 cm, n = 1,070) (Figure 3.7).

The vast majority of trees (86.7%) hosted lianas, of which 68.3% supported between 1 and 3 woody climbers (mean = 3, median = 2, n = 1,070). Albeit it was difficult to determine exact numbers, the rough breakdown of lianas

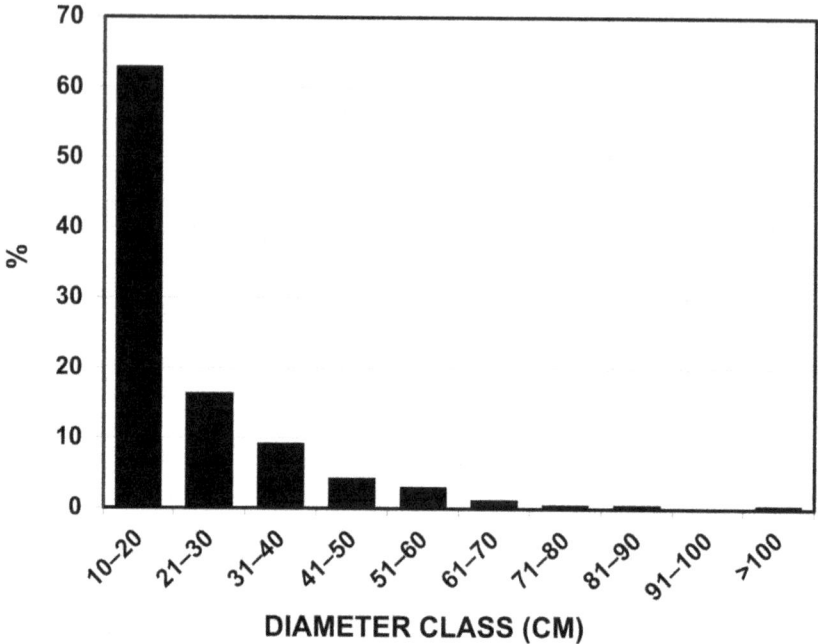

Figure 3.7 Diameter at breast height (DBH) of transect trees.

hosted amongst 948 trees was as follows: $1=24.2\%$, $2=27.7\%$, $3=26.9\%$, $4=15.4\%$, $5=3.6\%$, $6=1.6\%$, $7=0.2\%$, $8=0.3\%$, $9=0\%$, $10=0.1\%$. One would expect that thicker (and therefore typically larger) trees host more lianas, but no significant correlation was detected ($R^2=0.14679$).

At the end of the 2012–14 monitoring period, in Mar14, 28 trees had fallen to the ground, i.e. 2.6% of all transect specimens. These fallen trees hosted between 3 and 5 lianas. This average load of 3.6 woody climbers is slightly higher than the average load along the transect, indicating that the corresponding greater weight might have made the trees more vulnerable. In addition, signs of elephant activity on some fallen trees suggest damage by pachyderms.

Liana densities on Rubondo are very high, relative to other *Pan* study sites where comparable data have been collected; e.g. Gashaka, Nigeria, mean = 0.8 / tree, range 0–4; LuiKotale, DR Congo (bonobo study site), mean 1.8 / tree, range 0–6 (Sommer et al. 2011). Similarly, the 2001–03 study established that liana density on Rubondo is between four and nine times higher than comparable measures from three different Ugandan study sites (Moscovice et al. 2010). Akin to other sites where lianas proliferate, Rubondo has a relatively dry climate, and many tree fall gaps within the forest (Schnitzer & Carson 2001).

James Leonard – on Rubondo since 1998

Figure B3.2 James Leonard (r.) with Josephine N. Msindai (l.). (Photo: JNM)

I was in my teens when a friend of mine told me there were some *mzungus* looking for people to help them with research on chimpanzees. I had never heard of Rubondo and had certainly never seen apes. But I decided that it might well be interesting to work on the island. In those early years, life was tough; we slept in tents in the forest and had to cook our own food after long hours in the field. Over the years, I have worked with researchers who planned to habituate the chimpanzees, studied spotted-necked otters, trees and other plants; so, I have learnt a lot about wildlife and botany. Also, I sometimes had requests to collect chimpanzee hairs, not only for the researchers, but also for people on the mainland who wanted the hairs for black magic. One day, I saw three or four chimpanzees on a hilltop and started to track them. I passed a ground nest and followed them into a dense jumble of lianas. I suddenly heard their loud cries and realised I was really close to them. Suddenly, they were all around me, behind me, in front, and others that I could not see. A few of the chimps had babies. To me it seemed they were angry and upset at this intruder, perhaps they thought I was a threat to their young ones. I was very frightened, being in the middle of a group. I fell to the ground and crawled on my hands and knees slowly backwards through the thick mass of lianas. This was one of the few times I tried to get as far away as possible from the chimpanzees.

Tree and Liana Species

In the 2001–03 study, identification to species level was achieved for 99% of transect trees. The list consists of 53 species belonging to 32 families, of which 45.2% occurred in one region only, and only 24.5% in all three regions. In the 2012–14 study, identification to species level was achieved for 82% of transect trees. The list consists of 37 species belonging to 23 different families (Table 3.2, Figure 3.8). (Given that identification of tropical trees is notoriously difficult and their classification subject to constant revision, there are conflicting opinions about some of the names.)

The comparison made evident that the forest composition differs between parts of the island. On this account, the Kasenya region is heavily dominated by *Craterispermum schweinfurthii*, and also contains more *Croton sylvaticus* and *Vitex doniana* than the other regions. The Nyakutukula sector is dominated by *Drypetes gerrardii*, which is rare in northern regions. *Pancovia turbinata* and *Dichapetalum stuhlmannii* are common in the Kamea and Masakela sector, as well as the Nyakutukula sector, but not in Kasenya.

Despite these disparities between localities, the two surveys arrived at considerable overlap with respect to the most frequent tree species. Thus, the combined top 10 of both surveys is made up of only 14 different species, which constituted 82% of all specimens in the 2001–03 survey resp. 69% in the 2012–14 survey. In the following breakdown, the first number is the average percentage across both surveys, and numbers in brackets represent percentages during the 2001–03 resp. 2012–14 sampling periods:

- *Pancovia turbinata* 15.9 (14.7, 17.2);
- *Croton sylvaticus* 7.9 (10.1, 5.8);
- *Dichapetalum stuhlmannii* 7.7 (2.5, 13.0);
- *Alchornea hirtella* resp. *A. laxiflora* 6.1 (4.7, 7.6);
- *Teclea nobilis* 6.1 (9.1, 3.1);
- *Craterispermum schweinfurthii* 5.3 (9.8, 0.8);
- *Antiaris toxicaria* 4.3 (1.1, 7.4);
- *Synsepalum brevipes* 4.2 (4.3, 4.2);
- *Drypetes gerrardii* 4.2 (8.2, 0.2);
- *Haplocoelum foliolosum* 3.9 (6.5, 1.4);
- *Vitex doniana* 3.5 (4.4, 2.6);
- *Maerua duchesnei* 2.7 (1.8, 3.6);
- *Diospyros abyssinica* 2.1 (4.2, 0.0);
- *Carpolobia conradsiana* 1.3 (0.4, 2.2).

Chimpanzees consume parts from almost three-quarters of transect species – a fact we will revisit later (Chapter 5).

From 2012–14, only a few types of woody climbers could be identified, and some only to genus level, due to high degrees of similarity in stem and leaf morphology. However, the 2001–03 study identified 16 species (Table 3.3). Accordingly, two species, *Saba comorensis* and *Strychnos lucens*, together

Table 3.2 Tree species growing along two 4-km straight-line transects on Rubondo, in Kamea and Masakela areas. Established 2012; 2 m wide; 1,094 trees and 640 associated lianas, with 195 trees not identified

Name					Transect specimen	
Family	Latin	Common	KiSwahili	Source (a)	n	%
Sapindaceae	*Pancovia turbinata*	Goat-killer			188	17.2
Dichapetalaceae	*Dichapetalum stuhlmannii*	Christmas bush, Zulu bead-string		6	142	13.0
Euphorbiaceae	*Alchornea hirtella*			7	83	7.6
Moraceae	*Antiaris toxicaria*	Bark cloth tree, Antiaris, sacking tree	Mkunde	6, 7	81	7.4
Euphorbiaceae	*Croton sylvaticus*	Forest fever-berry	Msinduzi	2, 7	63	5.8
Sapotaceae	*Synsepalum brevipes*		Mchocha mke, mchocho jike	7	46	4.2
Capparaceae	*Maerua duchesnei*	Common bush-cherry			39	3.6
Rutaceae	*Teclea nobilis*			6	34	3.1
Verbenaceae	*Vitex doniana*	Black plum	Mfudu, mfuru, mfuu	7	28	2.6
Polygalaceae	*Carpolobia conradsiana*				24	2.2
Rutaceae	*Zanthoxylum chalybeum*	Knobwood, crocodile tree	Mjafari	7	24	2.2
Rubiaceae	*Rothmannia urcelliformis*			6	20	1.8
Sapindaceae	*Haplocoelum foliolosum*	Northern galla-plum		4, 7	15	1.4
Rubiaceae	*Canthium lactescens*	Hairy turkey berry		2	13	1.2
Guttiferae	*Garcinia huillensis*	Granite mangosteen			12	1.1
Mimosoideae	*Albizia gummifera*	Peacock flower	Mkenge, mchani mbao	7	10	0.9
Moraceae	*Morus mesozygia*	Black mulberry		6, 7	10	0.9
Rubiaceae	*Craterispermum schweinfurthii*			9	9	0.8
Anacardiaceae	*Pseudospondias microcarpa*	African grape		1	9	0.8
Moraceae	*Ficus sansibarica*	Fig tree			8	0.7

(*Continued*)

Table 3.2 Continued

Name		Common	KiSwahili	Source (a)	Transect specimen	
Family	Latin				n	%
Annonaceae	*Annona senegalensis*	Wild custard apple			7	0.6
Rubiaceae	*Coffea eugenioides*	Nandi coffee			7	0.6
Lecaniodiscus	*Lecaniodiscus fraxinifolius*	River-litchi	Mbwewe, mkunguma	4	7	0.6
Ulmaceae	*Celtis africana*	White stinkwood	Mbelangoma	7	6	0.5
Verbenaceae	*Premna angolensis*			7	4	0.4
Sapindaceae	*Zanka golungensis*				4	0.4
Phyllanthaceae	*Antidesma venosum*	Tassel berry	Mpungalira, mbua nono	2	2	0.2
Putranjivaceae	*Drypetes gerrardii*	Bastard white ironwood	Kihambie	4, 7	2	0.2
Sapotaceae	*Mimusops kummel*	Red milkwood	Mgambo	7	2	0.2
Loganiaceae	*Strychnos lucens*			4, 7	2	0.2
Rutaceae	*Zanthoxylum gilletii*	Satinwood, white mahogany		7	2	0.2
Combretaceae	*Combretum molle*	Velvet bush willow	Kagua, mdama, mlamam	3, 7	1	0.1
Ebenaceae	*Diospyros mespiliformis*	African ebony, jackal-berry	Mgiriti, mjoho, mgombe	2, 7	1	0.1
Flacourtiaceae	*Flacourtia indica*	Batoka plum	Duruma, madungatundu		1	0.1
Malvaceae	*Grewia forbesii mast*			5	1	0.1
Anacardiaceae	*Lannea fulva*				1	0.1
Rhamnaceae	*Maesopsis eminii*	Musizi, umbrella tree	Msizi, muhunya, ndunga	6	1	0.1

(a) Internet sites to arrive at tree names [accessed 2015-Apr-01–10]: (1) aluka.org, (2) plantzafrica.com, (3) worldagroforestrycentre.org, (4) zimbabweflora .co.zw, (5) theplantlist.org (6) wikipedia.org/en, (7) www.prota4u.org

Figure 3.8 Fruits and flowers of transect plants on Rubondo. T = tree, L = liana. (a) *Vitex doniana* (T); (b) *Rothmannia urcelliformis* (T); (c) *Carpolobia conradsiana* (T); (d) *Saba comorensis* (L); (e) *Ficus* sp. (T); (f) *Uvaria* sp. (L). (Photos: JNM)

Table 3.3 Liana species ≥1 cm DBH supported by trees ≥10 cm DBH, based on
971 plants on 6 randomly selected quadrats, sorted by total number of
individuals recorded. Scientific names follow nomenclature of Beentje
(1994, 2002)

Family	Species	%
Apocynaceae	*Saba comorensis (*var. *comorensis,* var. *floreda)*	34.0
Loganiaceae	*Strychnos lucens*	21.0
Annonaceae	*Uvaria* sp.	11.8
Celastraceae	*Salacia* sp.	10.1
Capparidaceae	*Capparis* sp.	4.6
Dilleniaceae	*Tetracera potatoria*	4.6
Papilionaceae	*Dalbergia malangensis*	4.2
Icacinaceae	*Pyrenacantha sylvestris*	3.2
Apocynaceae	*Baissea major*	2.8
Connaraceae	*Agelaea setulosa*	1.3
Nyctaginaceae	*Jasminum dichotomum*	0.8
Annonaceae	*Artabotrys modestus*	0.4
Vitaceae	*Cissus quadrangularis*	0.4
Passifloraceae	*Adenia rumicifolia*	0.3
Tiliaceae	*Grewia flavescens*	0.2
Rubiaceae	*Keetia venosa*	0.1

Source: Adapted from Moscovice 2006: 113

with two morphospecies, *Uvaria sp.* and *Salacia sp.*, account for 70.3%
of the 1,063 lianas surveyed. Half of the species (n = 8) were present in all
regions surveyed, 25% (n = 4) were present in two regions and the remaining
25% (n = 4) were present in one region only. *Saba comorensis* was among
the top two dominant liana species in all three regions.

Flowering

Angiosperms – flowering plants – reproduce via pollen, an often yellow, fine
powdery substance of microscopic grains, which is discharged from the male
part of a flower and carried to a receptive female ovule, typically from another
plant. Rain forest trees are rarely wind-pollinated because their high species
diversity diminishes the chances for a specific male gamete to reach a corresponding female organ. Also, high humidity and rain dampens the pollen and
causes its 'powder' to stick together, hampering aerial movements. Moreover,
the dense canopy filters pollen out and often renders airflow very still, conditions not conducive to wind-assisted pollen transport (Turner 2001).
Pollination is more likely achieved via animals such as insects, birds, bats,
primates, that prey on the flowers (or nearby fruits) and carry the pollen to
other trees (Smith-Ramirez & Armesto 1994, Heymann 2011). Consequently,
flowering is often associated with leaf fall, as leafless trees make blossoms
more conspicuous and more easily accessible to pollinators. Therefore, in seasonal forests, flowering often coincides with the dry season (Janzen 1967).

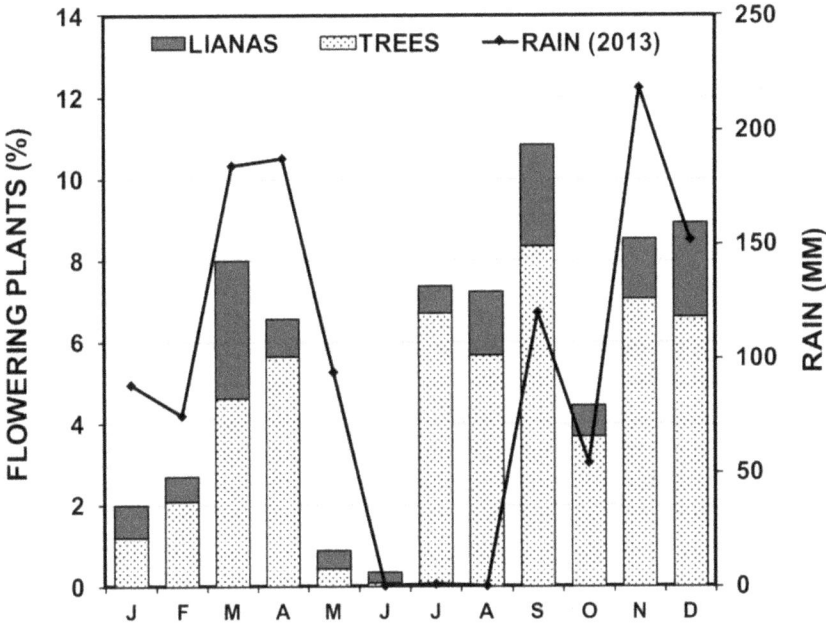

Figure 3.9 Flowering schedule of transect plants (Jan–Dec13) and annual rainfall.

On the 2012–14 Rubondo transects, on average, 10.4% of trees and 6.2% of lianas flowered each month, with, however, an uneven distribution throughout the year. Very few transect specimens blossomed in Jan–Feb and May–Jun, while all other months saw noticeable productions, with a peak in Sep (Figure 3.9).

Contrary to expectation, we were unable to detect any significant negative association of flowering and rainfall. In fact, there was a non-significant trend of more lianas flowering during months with higher rainfall (Pearson, $r=0.510$, $n=12$, $p=0.09$), while there was no such effect for tree flowers (Pearson, $r=0.338$, $n=12$, $p=0.3$). Similarly, rainfall 1 month prior to flowering had no influence on the percentage of lianas flowering the next month (Pearson, $r=-0.244$, $n=12$, $p=0.4$), albeit there was a non-significant trend of trees bearing more flowers when the previous month saw little rain (Pearson, $r=-0.510$, $n=12$, $p=0.09$).

These findings ascertain that plants can flower (or fruit, see below) during any time and that phenophases may vary greatly, even between closely related forms inhabiting the same forest. Such inter- and intra-specific variation is particularly expressed in African tropical forests, rendering flowering of forest trees notoriously unpredictable (Adamescu et al. 2018). Perhaps more importantly, except for a lull in rains from Jun–Aug, the climate on Rubondo is probably not seasonal enough to rigidly influence plant phenology (cf. Figure 3.3).

Fruiting

Once a flower is pollinated its petals wilt and the seed develops from the ovule. The ovary slowly grows into a fruit, which contains the seeds (Tutin & Fernandez 1993). Seeding can simply happen through fruits falling to the forest floor. However, if seeds are of a size and shape that can be dispersed by wind, then they ripen more likely during the dry season while air currents are stronger and branches contain fewer leaves, which enables greater dispersal distances (Chapman et al. 2005).

Dispersal may also be facilitated by animals attracted to the fleshy consumable parts of fruits. Unripe fruits are normally discretely coloured and generally dry, hard and bitter, so that they are more attractive for frugivores once the seeds have matured. Frugivorous animals tend to select brightly coloured fruit, often with red or black succulent pulp that stands out against a background of green foliage. Animals either discard undesirable seeds or ingest them with the pulp, to then excrete these indigestible parts (Turner 2001). Apart from birds, primates play an important role as seed dispersers in forests, as they can access canopy fruit, and then spit out or defecate undamaged seeds (Chapman et al. 2017). The more heavily plants rely on seed dispersers, the more likely is it that fruit ripens when conditions for dispersal are favourable. Moreover, the peak of production for animal-dispersed fleshy fruits is likely to coincide with the rainy season, probably because of such fruit depends on increased moisture levels (Chapman et al. 2005).

Two measures of fruit abundance on the 2012–14 Rubondo transects were used as proxies: the percentage of fruiting plants and the sum of the DBH of trees that bear fruits or host fruit-bearing lianas (fruit index). Mirroring the flowering pattern on the island, fruit was produced all year round. In any given month on average 10.2% of trees and 6.3% of lianas were fruiting (Figure 3.10). The fruit index closely followed the pattern of absolute numbers of fruiting plants, with peaks in Mar–Apr and Dec (Figure 3.11).

Overall, liana fruiting was less seasonal than tree fruiting. This is consistent with measurements during the 2001–03 surveys, which also established that 75% of the liana species on the island yield large fleshy fruits. In terms of their less pronounced reproductive seasonality, lianas benefit from adaptations for accessing and storing water, including a deep root structure and efficient vascular system (Opler et al. 1992).

Moreover, during the 2012–14 surveys, higher fruit production by liana plants was inversely proportional to tree fruit production in May, Aug and Jan, and when examined across the year came close to statistical significance (Pearson, n = 12, r = −0.484, p = 0.1). Lianas compete with trees for resources such as light, water and nutrients (Pérez-Salicrup & Barker 2000). Consequently they can reduce tree fruiting (Stevens 1987) and/or woody climbers might exploit lulls in tree fruit production so as to maximise the chance that their seeds will be dispersed.

There is some indication for the mentioned effect that fruit ripening coincides with more rains. Accordingly, the percentages of fruiting

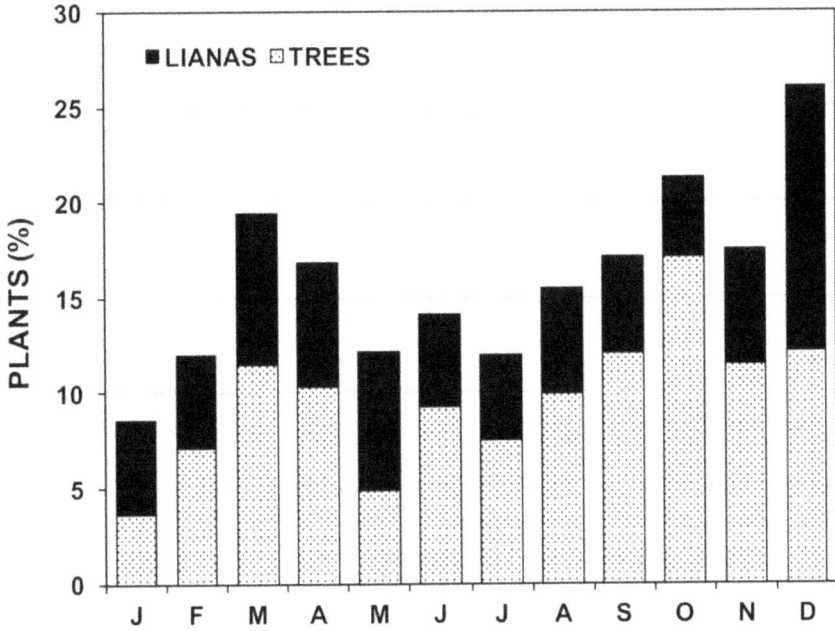

Figure 3.10 Fruit-bearing transect plants (Jan–Dec13).

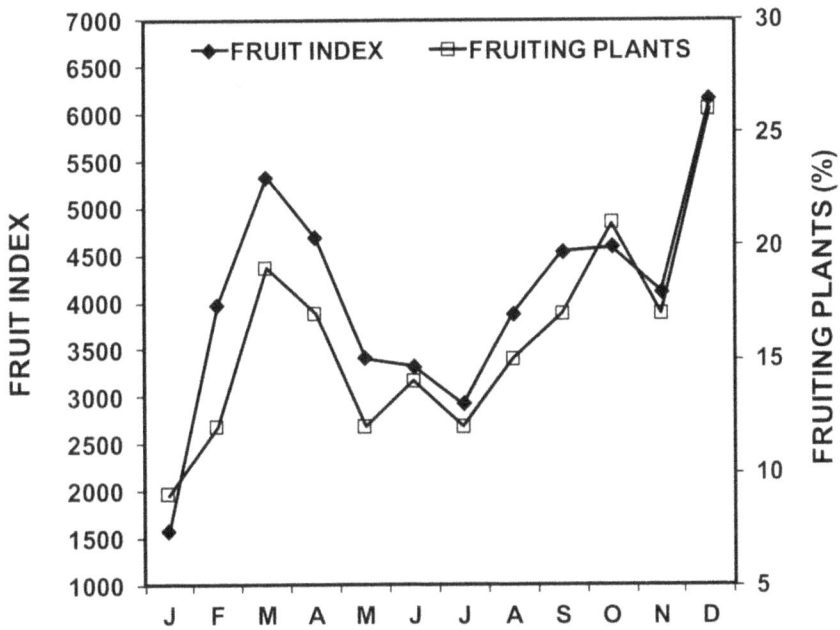

Figure 3.11 Transect plant fruiting schedule (Jan–Dec13), expressed as a comparison between (i) fruit index and (ii) percentages of fruiting transect plants.

transect plants were positively correlated with rainfall (trees, $R^2 = 0.092$, Pearson $r = 0.926$, $n = 12$, $p < 0.001$; lianas, $R^2 = 0.383$, Pearson $r = 0.683$, $n = 12$, $p = 0.01$). Similarly, there was a positive trend in that the fruit index increased during periods of higher rainfall (Pearson $r = 0.504$, $n = 12$, $p = 0.1$). Measurements during 2001–03 likewise found that tree fruits were most abundant during the wettest period of the year. In other tropical lowland rainforests inhabited by chimpanzees, fruiting also peaks during or shortly after high rainfall (e.g. Budongo: Tweheyo & Babweteera 2007, Issa: Hernandez-Aguilar 2006, Kalinzu: Kagoro-Rugunda & Hashimoto 2015), while the dry season corresponds with lower fruit availability (Furuichi et al. 2001, Doran et al. 2002).

Throughout the year, both primates and non-primates fed on fruits, as indicated by the more or less equal proportions of feeding remains left by these animals (Figure 3.12). We might expect that remains are mostly found during months right after when fruit availability had peaked. However, while such association has been established for, e.g. the study site of Gashaka (Jesus 2020), there was no relation between the numbers of fruit-bearing plants and remains on Rubondo.

Be that as it may, the outstanding feature of the 2001–03 and 2012–14 surveys is the fact that there is little seasonal shortage of fruit production in the Rubondo forests.

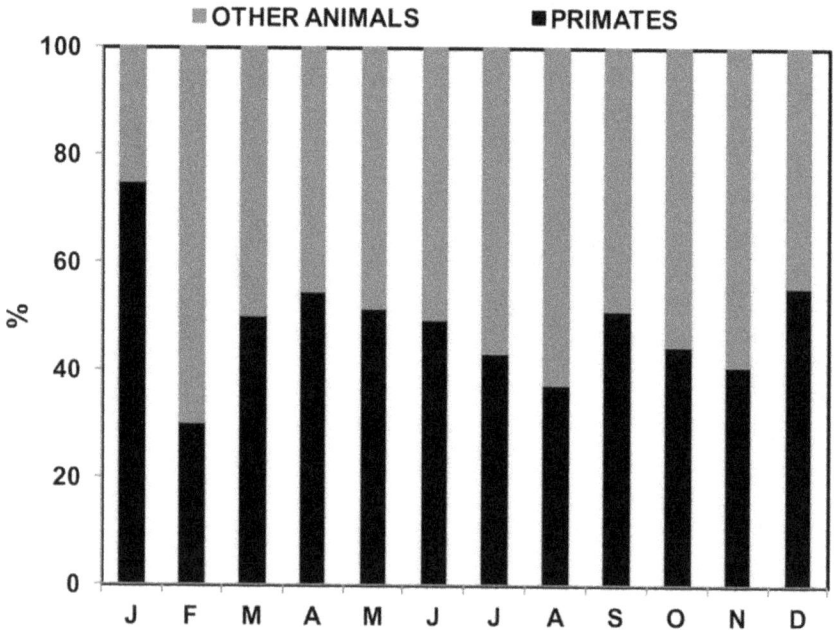

Figure 3.12 Proportions of primate and non-primate frugivores as assessed from feedings remains under transect trees.

Wildlife: Natives and Castaways

Apart from climate and vegetation cover, chimpanzees will also be influenced by other members of the faunal community, through factors such as resource competition, predation pressure, seed dispersal or disease vectors. On Rubondo, these effects are probably milder than in many habitats elsewhere, given an absence of other large-bodied non-human primates such as baboons or gorillas, as well as a lack of large predatory felids.

How will visitors conceptualise the overall embodiment of wildlife on Rubondo? Compared to places such as the Serengeti plains, tourists will certainly find it harder to spot animals (Figure 3.13). Still, even casual visitors

Figure 3.13 Large animals on Rubondo Island: (a) Nile crocodile; (b) hippopotamus; (c) guereza; (d) vervet monkey; (e) otter; (f) elephant; (g) sitatunga. (Photos: JNM)

who stay only by the shore can see otters, mongooses, monitor lizards, the fearless swamp dwelling sitatungas and bushbuck antelopes, plus the unmissable hippopotami and some enormous specimens of what is probably the largest population of crocodiles in Lake Victoria. The two types of monkeys, the endemic vervets and the introduced guerezas, can also be observed near the tourist lodgings and the shorelines at Kageya. Any organised 'game drive' will almost certainly yield sightings of introduced giraffes and elephants (cf. Table 1.1). Meanwhile, more than 100 pachyderms roam across the entire island including the shorelines (Mwambola et al. 2016). They also favour forest interiors, often in the same areas as chimpanzees (Chapter 4). However, given that the apes rarely leave the dense northern and central forests, they are difficult to spot for tourists, and certainly not from the back of a land rover. Consolation might be found during boating trips promising a haul of sizable tilapia or humongous Nile perch – of which the largest caught specimen weighed 184 kg (Figure 3.14). (These two other types of introduced animals, as we will discuss, had a devastating impact on the lake's underwater ecology).

Several field guides to East African wildlife (e.g. Withers & Hosking 2002, Stevenson & Fanshawe 2020, Kingdon 2021) detail the share found on Rubondo. Our brief compilation of the national park's larger-sized animals (Table 3.4) includes 15 mammals, 11 reptiles, 3 amphibians, 3 types of easily identifiable fish plus 29 bird species. However, bird sightings can tally more than 300 species from Dec–Mar, when European migrants visit the island.

A compelling intellectual justification to mention the following events is hard to come by, except that natural history anecdotes may become foundations for overarching theories. In this spirit we provide descriptions of some rather unusual encounters with wildlife – beyond the already described chimpanzee attacks (Chapter 2). Nothing short of spectacular was a 1994 observation by Guido Müller (Müller & Anzenberger 1995) of a fight beween an elephant bull and a hippo. The Swiss researcher witnessed the pachyderm shovelling the hippo – not yet fully grown, but still – on his tusks, and then flunging the adversary to its death. Of course, providence failed and no camera was at hand...

Along the theme of unusual encounters also ranks a Jul12 episode when this book's co-author was the only occupant in a beach lodge, courtesy of Asilia tourism company. The sturdy tent included a private bathroom and a kitchenette. Sometime during a good night's sleep, VS became semiconscious because he inhaled a deeply penetrating musky odour. Soon afterwards, he was awoken by what seemed to be an arm pushing through the mosquito-netting of the window. To his astonishment, it was a trunk of an elephant male in musth – the periodic condition of unpredictable behaviour associated with five, six times the normal testosterone levels and heightened aggression. The bull seemed determined to dismantle the tent's timber frame. The hapless occupant sought immediate refuge under the sink in the

Figure 3.14 Illegal and legal fishing on Rubondo. (a) Poachers with their catch apprehended by TANAPA rangers (1994). (Photo: Guido Müller) (b) Tourists with their catch (2012). (Photo: William Rutta, Bukoba Cross Culture Travel and Tours)

Table 3.4 Fauna of Rubondo Island National Park: selected vertebrate animals

Group	Family	Species	Common name
Mammals	Bovidae	*Neotragus moschatus**	Suni antelope
		Tragelaphus scriptus	Bushbuck
		Tragelaphus spekii	Sitatunga
	Cercopithecidae	*Chlorocebus aethiops*	Vervet monkey
		*Colobus guereza**	Guereza or black and white colobus monkey
	Elephantidae	*Loxodonta africana**	Elephant
	Giraffidae	*Giraffa camelopardalis**	Giraffe
	Herpestidae	*Atilax paludinosus*	Marsh mongoose
	Hippopotamidae	*Hippopotamus amphibius*	Hippopotamus
	Hominidae	*Pan troglodytes**	Chimpanzee
	Mustelidae	*Lutra maculicollis*	Spotted-necked otter
	Procaviidae	*Procavia capensis*	Rock hyrax
	Pteropodidae	*Epomophorus minimus*	Fruit bat
	Suidae	*Potamochoerus larvatus*	Bushpig
	Viverridae	*Genetta servalina*	Genet
Amphibians	Bufonidae	*Bufo regularis*	Toad
	Hyperoliidae	*Afrixalus brachycnemis*	Short-legged leaf-folding frog
		Hyperolius viridiflavus	Common reed frog
Reptiles	Chamaeleontidae	*Chamaeleo dilepis*	Flap-necked chameleon
	Colubridae	*Hapsidophrys smaragdina*	Emerald snake
		Philothamnus battersbyi	Battersby's green snake
	Crocodiles	*Crocodylus niloticus*	Crocodile
	Elapidae	*Dendroaspis jamesoni*	Jameson's mamba
		Dendroaspis polylepis	Black mamba
		Naja melanoleuca	Black forest cobra
	Eublepharidae	*Hemidactylus mabouia*	Common house gecko
	Pelomedusidae	*Pelusios williamsi*	William's hinged terrapin
	Pythonidae	*Python sebae*	Rock python
	Varanidae	*Varanus niloticus*	Nile monitor lizard
Birds	Accipitridae	*Haliaeetus vocifer*	African fish eagle
		Milvus migrans	Black kite
	Acrocephalidae	*Iduna pallida*	Olivaceous warbler
	Alcedinidae	*Alcedo cristatus*	Malachite kingfisher
		Ceryle rudis	Pied kingfisher
		Ispidina picta	Pigmy kingfisher
	Anatidae	*Alopochen aegyptiaca*	Aegyptian goose
	Ardeidae	*Ardea purpurea*	Purple heron
		Ardeola ralloides	Squacco heron
		Egretta garzetta	Little egret

(Continued)

Table 3.4 Continued

Group	Family	Species	Common name
	Bucerotidae	*Bycanistes subcylindricus*	Grey-cheeked hornbill
	Burhinidae	*Burhinus vermiculatus*	Water thicknee
	Charadriidae	*Vanellus senegallus*	African wattled lapwing
	Ciconiidae	*Anastomus lamelligerus*	Open-billed stork
	Estrildidae	*Lagonosticta senegala*	Red-billed firefinch
	Jacanidae	*Actophilornis africanus*	African jacana
	Meropidae	*Merops albicollis*	White-throated bee-eater
		Merops pusillus	Little bee-eater
	Monarchidae	*Terpsiphone viridis*	African paradise-flycatcher
	Motacillidae	*Motacilla aguimp*	African pied wagtail
	Musophagidae	*Crinifer zonurus*	Eastern plantain-eater
	Phalacrocoracidae	*Microcarbo africanus*	Long-tailed cormorant
		Phalacrocorax lucidus	White-breasted cormorant
	Ploceidae	*Amblyospiza albifrons*	Grossbeak weaver
		Euplectes progne	Long-taileded widow
		Ploceus nigerrimus	Vieillot's black weaver
	Psittacidae	*Psittacus erithacus**	African grey parrot
	Scolopacidae	*Actitis hypoleucos*	Common sandpiper
		Tringa glareola	Wood sandpiper
Fish	Latidae	*Lates niloticus**	Nile perch
	Cichlidae	*Oreochromis niloticus**	Nile tilapia
		100s of species	Haplochromine cichlids

* = introduced (see Table 1.1)
(Records by JNM and TANAPA)

kitchenette, but when doing so, hit his toe against a tiled corner. While the digit began to bleed profusely, the pachyderm continued his rampage in a fashion that prophesied ultimate demise. Fortunately, after what seemed like eternity, the bull turned his attention to the surrounding ornamental palm trees, which he ripped out and flailed. Throughout much of the next day, the mighty bull hung around camp in a distinctly menacing demeanour. VS later described the incident to his son, who sent a befitting YouTube link to Maceo Parker's funk-piece 'Elephant stepped on my toe'. Indeed, it took a year for the toenail to grow back.

The main author of this book (JNM) had her own share of an encounter too close for comfort in Nov13. In the afternoon, she set off back to camp on her motorbike after a day in the forest. Unfortunately, en route, a bush-buck ran out of the thicket, probably disturbed by the noise of the engine, and collided with the bike. JNM went flying off, dislocated her elbow and broke her forearm's radius bone. Tourist camp managers, alerted via radio, rushed to collect her with a landrover, but couldn't reach because of a fallen tree. For this reason, Samson Obiedy, one of the trackers, came to the rescue with his motorbike. The evacuation air insurance provider refused help, alleging it dangerous to land an aircraft on the island, even though commercial local planes regularly used the airstrip. The park management stepped in to help and organised a boat. By this time, it was after sunset, with lake waters somewhat choppy. After an hour on treacherous waves, the boat reached the mainland ranger post at Nkome. From here, a car ferried JNM plus entourage to the gold-mine town of Geita, arriving at midnight. The rangers had called ahead and a kind nurse had stayed on at the hospital, providing an X-ray and stabilising the wrist. As the hospital's dispensary was closed, pain medication had to be acquired from a nearby private clinic – including several vials of Valium to be injected into the gluteal muscles back at a hotel. Follow-up surgeries were undergone in Dar es Salaam and London. Even after years of physiotherapy, the event is painfully remembered on a daily basis.

Apart from such accidental collisions, Rubondo Island National Park, as we have mentioned, comes across as a surprisingly peaceful wildlife sanctuary. However, the impression is misleading, as the tranquillity disappears, once we delve into those 200 km^2 that are not on dry land – meaning, the world underwater. Here, like in the terrestrial parts, introduced animals are major players, only that this time round their release has caused an ecological catastrophe.

To biologists, Lake Victoria is known as 'Darwin's Dreampond', because – like Galapagos with its finches or Hawaii with its honeycreepers – it saw an astonishing adaptive radiation, in this case of the perch-like cichlid fish that evolved into hundreds of closely related but morphologically variant species, distinguishable by diverse sizes, shapes, colours and feeding strategies (Goldschmidt 1996).

In the early 1900s the lake had a multi-species fishery industry centred on two endemic cichlids, the tilapia *Oreochromis esculentus* and *O. variabilis*. Fish exploitation increased following the extension of railway lines and the introduction of more efficient gill nets in 1905. A 1928 survey highlighted that the tilapia was over-fished and recommended restrictions on the use of gill nets. However, no limits were ever put in place. Thus, within a decade the fishery collapsed, with *O. esculentus* becoming an endangered species. In efforts to boost commercial fishing, the British colonial administration introduced Nile tilapia (*O. niloticus*) to Lake Victoria sometime in the 1940s (Ogutu-Ohwayo 1990). This fish can reach 60 cm in length and exceed 5 kg. To further improve lake productivity, proposals in 1929, 1932 and again in 1959 recommended the introduction of a commercial predator to convert the hundreds of small and low-value cichlid species into a more economically profitable entity. Concerns about the biological implications of such an interference (Anderson 1961) were pushed aside, and in the early

1960s, one of the largest predatory freshwater fishes was additionally introduced: the Nile perch (*Lates niloticus*), which can grow up to nearly 2 m
in length and reach 200 kg in weight (Pringle 2005). Within a decade, its
numbers had noticeably increased and fishing became commercially viable.
Meanwhile, in 1973, the Dutch had brought trawlers from the Netherlands
to the lake "to collect 60 tons of cichlids per day" for processing into fish
meal (Goldschmidt 1996: 4). By the 1980s, of the lake's original 500 species
of haplochromine cichlids, which had previously accounted for 80–90% of
species, only roughly 200 would remain (Ogutu-Ohwayo 1990). The lake
came to be dominated by three species, given that they were increasingly
freed from food competition with endemic cichlids: the introduced Nile tilapia, known as Ngege in KiSwahili; introduced Nile perch, known as Mbuta;
and the native sardine-sized silver cyprinid (*Rastrineobola argentea*) from
the carp family, known as Dagaa.

As the number of Nile perch in particular swelled in the lake, a burgeoning fishing industry developed on its densely populated shores. At its
peak in the 1990s, Lake Victoria, with some 1.6 million tonnes of Nile
perch removed annually, was the single most important source of freshwater fish in the world. The environmental and social effects of this enter-
prise were the subject of the 2004 Austrian–French–Belgian film *Darwin's
Nightmare*. The controversial documentary by Hubert Sauper shows
planes from Europe landing on Mwanza airfield that will then fly out pro-
cessed fillets of Nile perch – while locals are left to consume the festering
carcasses of the gutted fish. Eventually, the film was banned in Tanzania
because of the perceived negative light in which it portrayed the lives of
people primarily in Mwanza. The staff of the NGO which helped the film-
makers were either jailed or deported by the then president. In the West,
the film received acclaim, because it highlighted how foreign consumerism
is enabled by the misery and suffering of local citizens and the destruction
of their environment.

Nile perch numbers fell in the 2000s, and only about 800,000 tonnes were
removed in 2015 – still enough to sustain a substantial fisheries industry.
The decline is probably the unavoidable consequence of overfishing (Taabu-
Munyaho et al. 2016). In addition, huge swathes of wetlands at the shores
were converted into farmland, which removed important breeding grounds
for fish (Obiero et al. 2012). Moreover, the influx of nutrients from deforestation, crop production and general human pollution resulted in eutrophication of the lake, nourishing toxic algal blooms. The likely chain of events
leading to this disastrous situation began with the Nile perch introductions,
which destroyed the haplochromine population, inevitably disrupting food
chains and nutrient cycling, which in turn favoured the lake's eutrophication
(Marshall 2018). Within this context, Rubondo and its wetlands still serve as
important refuges for breeding fish. Its surrounding waters – as the the only
protected parts of the lake – continue to support a sizeable stock of Nile perch.

In short, both sections of Rubondo Island National Park, the one above the waters and the one below, are heavily affected by introductions of non-endemic animals.

Human Impact: Rangers, Poachers, Tourists

Since ancient times, diverse human populations of foragers have tracked the shrinking and expanding shoreline of Lake Victoria and associated herbivore populations (Tryon et al. 2016). Today, the lake's surroundings are some of the most densely settled areas in sub-Saharan Africa (Nelson 2004), not least, because the 1980s saw a huge influx of migrants from neighbouring DR Congo and those fleeing genocide in Rwanda. The Rubondo vicinity is likewise heavily populated. Nearby Mwanza is the second largest city in Tanzania, and the region counts 2.8 million people (UNFPA 2012). Still, almost miraculously, Rubondo is literally an island of peace, virtually cut off from human hustle and bustle – and protected by the fact that it was, back in 1977, declared a national park.

In the northern forests, one comes across old logging camps. However, these were abandoned many years prior, and there is no indication of recent logging. In fact, it is hard to see how economically profitable amounts of wood could be ferried away in small boats.

Still, people from the mainland and nearby islands will come to Rubondo to hunt for antelopes – bushbuck or sitatunga, also occasionally the rare suni. Poachers seem to concentrate on the central and southern parts of Nyakutukula and Lukaya, where wires and pit traps were discovered by the main author in Jul13. Around the same time, park rangers caught hunters with a large horde of antelope meat. Interestingly, while a fair few bush pigs roam the landscape, there were no reports that they were being hunted. In fact, park rangers lamented to JNM that they have never tasted the meat of wild pig from the forest.

Two species of introduced animals – rhinoceroses and roan antelopes – were hunted to extinction within a decade or so of them being transferred to the island (Chapter 1). Their final demise happened when Rubondo was already a national park. The introduced elephants, despite their prized tusks and some incidents of poaching (Wickson Kibasa, pers. comm. to JNM), were spared the fate of extinction. The first author, in 2012, in one open clearing in Kamea, counted 22 animals, but, in the same year, also heard gun shots and the hollering of elephants – so, poaching likely still occurs, however sporadic.

A report from 1973 on the then 'Rubondo Island Game Reserve' provides some flavour about the difficulties of anti-poaching measures before the place was declared a national park (Matschke 1974). In it, the German forestry volunteer Wolfgang Matschke complains that "an island by itself, worst still covered with dense forest" cannot be sufficiently guarded by 11 allocated employees ("1 Game Assistant, 5 Field Assistants, 5 Baggage Attendants"). In addition, "at least 50% of the usual working time was unproductive and spent outside the island because of sickness (own or of family members), leave, court summons, private business etc.". Consequently, while there were camps in the forest, "only one poacher was caught", plus "40 fishermen illegal[ly] fishing along the shore were arrested and taken to Geita police station. One little motor vessel, about 20 canoes and appr. 150 fishnets were confiscated." He concludes: "If eleven people stay in 5 camps and often even outside the island, how can proper patrols be possible?", urging "all concerned authorities not to disregard the problem again". While Matschke points out prevailing struggles, he equally praises the willingness of "third-world nations" to develop nature reserves in the first place – adding that this determination puts European nations and particularly his home country Germany "at shame" (Matschke 1976: 26) – echoing Bernhard Grzimek's sentiments (Chapter 1).

As is evident from such descriptions, the main illegal activity is fishing, as people are lured to the park's waters which still contain a fairly large stock of Nile perch and Nile tilapia. Guido Müller (pers. comm.), during his research in 1994, was invited by rangers to join an anti-poaching sting. Law-breaking fishermen hide their crafts and themselves in the wide reed belts at forested parts of the shore. A couple of days earlier, rangers had discovered and seized one such vessel, albeit the poachers had not been apprehended. This left the illegal intruders with only the option to escape on a raft. The rangers cruised along the eastern coast where, indeed, they spotted two men paddling a scanty structure. The steersman throttled engines, explaining that the officers would let them paddle for a while, to tire them out. Ultimately, the duo was apprehended without aggression on either side, law-breakers and law-enforcers displaying a matter-of-fact attitude. During their stay on the island for half a week they were subjected to manual labour before being taken to court in Geita. The confiscated catch was smoked and brought to the mainland as evidence in the trial. The rangers were not allowed to keep any of the fish, so as to prevent them from selling the loot for profit.

Similarly, once Müller led rangers to the hippo that had been killed by an elephant (see above), the rangers, instead of conserving its meat, burnt the entire animal. However, other sources maintain that rangers burn confiscated animals primarily, when the meat is already rotten or if there is danger that poachers poisoned their quarry before fleeing the scene. Nowadays it also seems established practice that confiscated fish are smoked and then shared between park employees and their families. On occasion, the confiscated horde is so large that some parts are sent to ranger families on the mainland as '*zawadi*' – gifts. During our research from 2012–14, fishing camps were still encountered on the shores in most regions of the island. Sometimes, the fishermen would run away, at other times they were friendly. On one occasion, the research team spotted 8–10 boats which remained there for an entire week. These crews appear to be well organised, suggesting that wider schemes are in existence. At any rate, at the nearest mainland village Mganza, trucks filled with fish depart regularly for neighbouring Uganda and Burundi. In informal discussions, some fishermen revealed that they visit the island frequently. One told that they often hear chimpanzee calls when camping out in the forest.

Roughly 1,200 fishermen per annum are caught and jailed for illegally fishing in Rubondo's waters (Fredrick Chuwa, TANAPA warden for protection, pers. comm. to JNM). The maximum punishment for illegal fishing is a 10-year prison sentence. However, a detainee in a holding cell expressed confidence he would return within a few months. Many fishermen only serve a 1-month prison term, because of corruption in the criminal courts on the mainland.

The archetypical divide between rich and poor at the heart of this dilemma (c.f. Figure 3.14) is embodied by the fact, that fishing is not illegal, if conducted by tourists, provided they buy a daily license for 50 USD (foreigners) resp. 5 USD (Tanzanian). In 'sports fishing', the catch will be measured and released, but mostly, the fish are consumed by the tourists, park employees and researchers.

Finally, on the topic of tourism – what about Bernhard Grzimek's sales pitch to the Tanzanian government, that a national park on Rubondo would create revenue (Chapter 1)? In 2018–19, Tanzanian parks were visited by 731,351 foreign and 464,933 domestic visitors. Of the foreigners, 80% primarily visited the three parks Serengeti, Kilimanjaro and Ngorongoro crater (TANAPA 2020). In contrast, Mwanza and other nearby towns feed only a small stream of visitors that come for recreation to Rubondo. In 2013, this amounted to 700 people, one-third of them Tanzanian, and visitors mainly partake in fishing excursions.

Thus, the aspirations of Grzimek and Archard of turning the island into a popular midway stop between more prominent travel destinations never materialised. Overall, and compared to other Tanzanian national parks, Rubondo remains one of the least visited. There is but a trickle of travellers, including a minority of well-heeled clientele indulging in luxury safaris. The licensed private tourism companies – primarily Asilia (see asiliaafrica.com) – take great care to not damage the island's treasures they want to sell; in fact, they do their best to supplement and support the government's efforts to protect nature and wildlife.

All said and done, Rubondo is certainly not a park where nature 'pays for herself'. Tourism revenue does not offset the expenditure of managing the reserve, constituting a bare 15% of the operational costs (Kevin Nkulila, chief park warden for tourism, pers. comm. to JNM). To be cost-effective, about 400 tourists per month would be needed. Such numbers are not even seen in Gombe Stream National Park, despite its worldwide fame due to Jane Goodall's work there with the chimpanzees; Gombe counts just 1,700 visitors per year. On the other hand, it would be detrimental if tourism numbers on Rubondo would increase in conjunction with chimpanzee viewing – given that a few dozen animals would be pursued by about 15 people per day. The alternative is to raise the price of the fee – which, in fact, Rubondo has recently implemented, charging 125 USD for the hope of seeing the animals during a newly launched 'Wild Chimpanzee Habituation Experience'.

In sum, Rubondo creates a financial deficit. This minus means that biodiversity conservation comes at a price. But, collateral benefits may be more difficult to measure – such as the reputational gain for Tanzania.

4 Bound to Be Wild: Sociality and Ranging

Figure 4.1 Rubondo chimpanzee female. (Photo: JNM)

DOI: 10.4324/9780367822781-4

Their ancestors were snatched from their wild-living mothers as infants and forced into a captive life to entertain humans, to then be tossed back into the wild – a chain of events that surely caused deep emotional distress. We might therefore wonder if the Rubondo apes are 'normal' chimpanzees (Figure 4.1). That is the question we will reflect on in this chapter. For this, we assess the degree to which the introduced apes have adopted species-typical patterns of gregariousness by exploring similarities with and differences from native populations.

The main features of sociality we investigate are the chimpanzee-typical dynamics of fission–fusion by which a community splits into smaller parties, with these units meeting and separating again, while at times exchanging members. By investigating the pattern observed on Rubondo, we will try to understand which environmental factors cause them, and if the spatial and temporal variation falls within the range of fluctuations seen in natural populations. In a similar vein, we use evidence for the presence of chimpanzees, such as sightings, tree and ground nests as well as other traces, to quantify the extent of the space they use and associated seasonal variation. By doing so, we compare our own data with those collected by previous researchers one to two decades ago, to discern notable changes.

The dimensions of fission–fusion and ranging routines will be influenced by the overall population size. Hence, we project how about a dozen surviving founder animals may have multiplied over the last half century to form today's population – and discuss if these descendants live in one community, or if they have formed competing units. In other words, we investigate the degree to which the Rubondo apes have rehabilitated themselves.

Chimpanzees: A Biosketch

For context, we first outline general biological characteristics of the genus *Pan*. Our knowledge about *Pan troglodytes* has, since the 1960s, continuously deepened through ever more detailed studies in the field, in captive settings and via molecular analyses – as evidenced by a string of often site-specific monographs (Goodall 1986, Boesch & Boesch-Achermann 2000, McGrew 2004, Reynolds 2005, Matsuzawa et al. 2011, Nishida 2011), and edited volumes (e.g. Wrangham et al. 1994, Boesch et al. 2002, Lonsdorf et al. 2010, Sommer & Ross 2011, Nakamura et al. 2015, Boesch & Wittig 2019). Consequently, there is a vast body of literature on chimpanzees. While a thorough review is not the aim of our book, we will nevertheless map out basic features of *Pan* phylogeny, distribution and socioecology (based on similar summaries in Pascual-Garrido 2011, Msindai 2018, Jesus 2020). This overview will provide a framework for our analyses of the gregariousness (Chapter 4), nest building and foraging behaviour (Chapter 5) of the Rubondo apes.

To contextualise our findings, we outline phylogenetic relationships amongst the various types of apes – acknowlediging that experts disagree on taxonomic terminology (e.g. Mann & Weiss 1996, Wildman et al. 2003, Arnold 2008). Within the order of primates, all apes belong to the superfamily Hominoidea. This taxon is divided into two families. The Hylobatidae ('small apes') include gibbons and siamang, restricted to South and Southeast Asia. The Hominidae ('great apes') include the Asian taxon orangutan as well as the African taxa gorilla, chimpanzee, bonobo, with humans also being part of that African clade. Genetic investigations established that chimpanzees and bonobos are less closely related to gorillas than they are to humans. This contradicts the superficial impression that the hairy knuckle-walkers (gorilla, chimpanzee, bonobo) form a separate group from the hairless upright-walking humans. Still, all four forms of ape stem from a common ancestor. Their paths split 8–9 million years ago, when one branch led to gorillas (tribus Gorillini) and a second branch to chimpanzees, bonobos and humans (tribus Hominini; Figure 4.2). From a last common ancestor that lived as recently as 4–6 million years ago, the Hominini again bifurcated into the subtribus Panina which contains the extant genus *Pan* (chimpanzee, bonobo), while the extant genus *Homo* forms the subtribus Hominina (other anthropologists label humans as 'hominids' or 'hominins'). Humans gradually fanned out, with only *Homo sapiens* surviving to the present day (extinct genera include *Paranthropus*, *Australopithecus*,

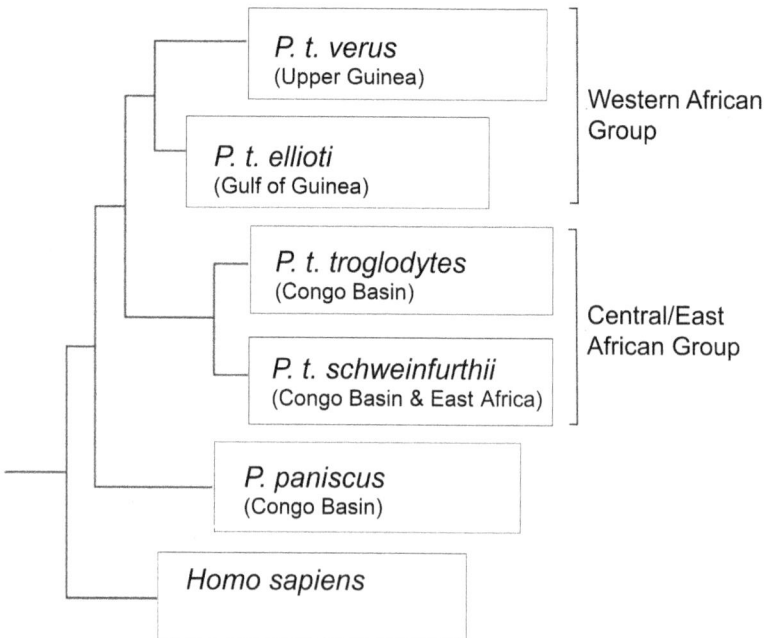

Figure 4.2 Phylogenetic relationships amongst the Hominini.

Ardipithecus). Their sister group, the Panina, split only 1–2 million years ago. Bonobos (*P. paniscus*) established themselves in lowland rainforests south of the Congo River. North of the riverbank, chimpanzees (*P. trog-lodytes*) began to populate diverse biotopes in altitudes of up to 3,000 m, including semi-deciduous and closed canopy rainforests, gallery forests, open woodland–savannah and mosaic habitats with grassland and – today – plantations.

Chimpanzees exhibit the widest geographical distribution of all apes due to their considerable ecological flexibility (Caldecott & Miles 2005). Still present in 21 African countries, they are extinct in Gambia, Benin, Burkina Faso, Togo and Zambia. They survive from Senegal in the north-west to the Congo in the south and Tanzania in the east, forming four rec-ognised subspecies. The genetic variability of chimpanzees is thus greater than that of bonobos on the left bank of the geographical barrier (Figure 4.3). Bonobos have black faces, with a naturally parting mop of hair. Chimpanzees look only slightly different according to subspecies. West African forms sport a relatively thin fur and dark face – hence known as the 'masked' chimpanzee – which has an even darker hue in the 'black-faced' Central African variety. The Nigeria–Cameroon variety does not seem to possess distinguishing characteristics, and can show either black or pale faces, while East African animals are distinguished as 'long-haired' chimpanzees. Although we will not generally extend our comparisons to bonobos, their basic sociality and ecology is similar to chimpanzees. Nevertheless, chimpanzee societies are male-dominated while bonobo societies are female-dominated, inspiring a distinction between 'patriar-chal' chimpanzees and 'matriarchal' bonobos (Sommer et al. 2011).

Chimpanzees are large-bodied primates, with an average adult male weight of 44 kg (range 37–60 kg) compared to 36 kg (range 30–47 kg) for adult females. They are agile arborealists, but can also rapidly knuckle-walk on the ground and may sometimes attain an upright position, to reach a food source or during displays. Chimpanzees form multimale–multifemale groups. These so-called 'communities' or 'unit-groups' typically include 30–80 members, but there is tremendous variation from just 8 members (Bossou, Guinea; Hockings et al. 2017) to 187 members (Ngogo, Uganda; Carlson et al. 2013). Communities typically range over 7–38 km^2 in forests and 50–560 km^2 in arid open habitats.

The patchy distribution of food causes individuals within communities to break up and forage in smaller units, typically comprising 6 members (range 3–10). Each night – and at times during the day – every group mem-ber (except dependent offspring) will build a new nest ('sleeping platform') from leafy twigs, typically in trees but sometimes also on the ground. The subgroups communicate with each other via long-distance acoustic signals, i.e. vocalisation (e.g. pant-hoots) and drumming (e.g. beating hands or feet on tree buttress roots). The composition of subgroups is not fixed. Instead, as time passes and animals move throughout the environment, individuals

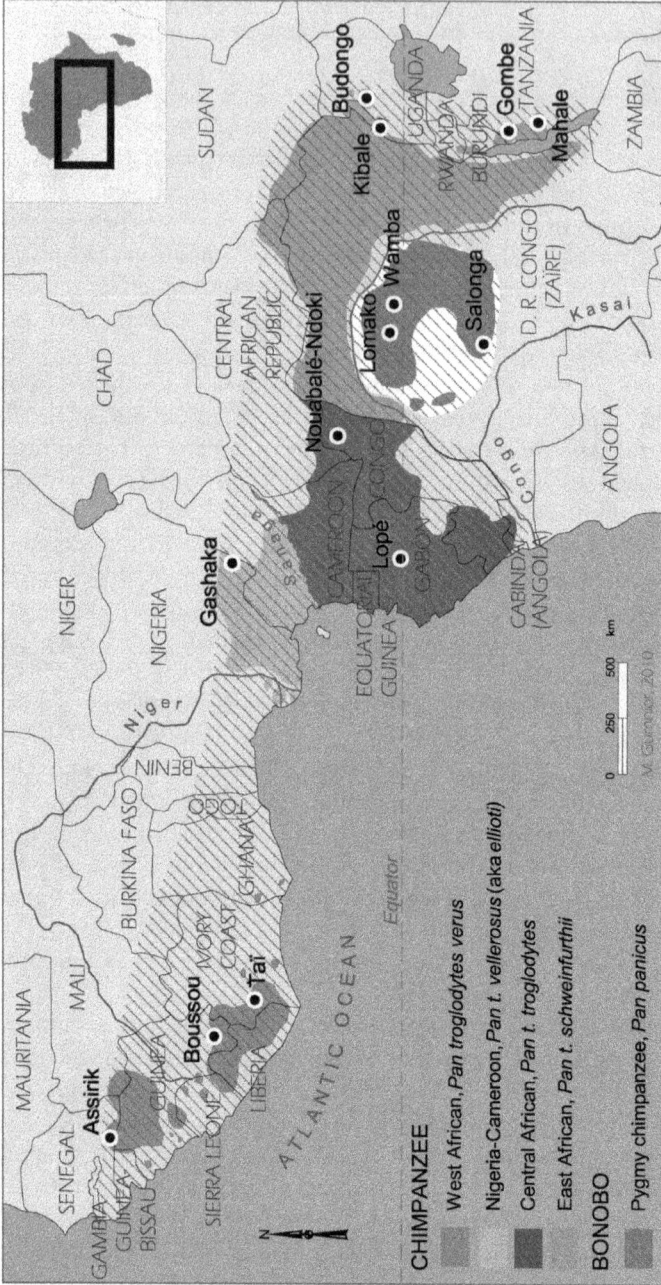

Figure 4.3 The genus *Pan*: Geographical distribution of chimpanzee subspecies and bonobos. Circles indicate selected study sites. (Map design: VS; drawing: Maren Gumnior)

may split off and join other units – a property embodied in the term 'fission–fusion society' (review in Newton-Fisher et al. 2000). Males remain in their natal community – they are philopatric – whereas females tend to emigrate upon sexual maturity into a neighbouring group. As a consequence, males are generally closely related and cooperate to defend their range against competing groups. Given that chimpanzees try to maintain exclusive access to certain parts of the landscape, home-ranges are called 'territories'. As a result, aggressive intergroup encounters are common and may go on for years until a neighbouring group is exterminated ('proto-warfare' or 'lethal raiding', Wrangham & Peterson 1996, Mitani et al. 2010, Wilson et al. 2014). Chimpanzee diet varies considerably across populations. However, fruit is always their main food, while they also consume leaves, flowers, bark or herbs. At some sites, the animals supplement their menu with social insects – eating ants, termites, honey – or they may catch mammals such as small forest antelopes or monkeys. At some places, they use stick tools or hammer tools for extractive foraging or in social contexts (e.g. Pascual-Garrido et al. 2012, Almeida-Warren et al. 2017).

Each chimpanzee community is characterised by a unique combination of social customs, tool-kits, communication, territorial aggression, hunting strategies, and plant consumption for food and self-medication. Thus, while belonging to the same species, members of particular groups, based on social transmission of information, might develop different technologies to solve environmental problems, as well as customs and traditions, creating local social identities. This diversity has become a staple theorem of 'cultural primatology', and is viewed as a homologue to 'cultural variation' in humans (Whiten et al. 1999, McGrew 2004, Sommer & Parish 2010).

Fission–Fusion: Splitting and Merging

Like many animals, primates tend to live socially, because being with others has advantages over living alone (review in Alcock 2009). In theory, for groups to form, the benefits of sociality must outweigh the costs. However, balancing the two means accepting that conspecifics can transmit pathogens and compete for resources such as food, resting space or mates. Moreover, foraging in groups necessitates longer distances and, with this, expending additional energy. On the flip side, sociality decreases predation pressure via the effect of dilution (safety in numbers), many-eyes (earlier detection, alarm) and deterrence (mobbing, defence) – main axioms of W.D. Hamilton's theory of 'geometry for the selfish herd' (Hamilton 1971). Also, individuals can cooperate and defend resources and mates from other competing groups, companions may transfer information about where to find food, where to expect danger, exchange useful skills and assist each other in raising young – up to the extreme of cooperative breeding.

The trade-off of costs versus benefits varies with ecological circumstances, producing diverse solutions for the animals in question, often tied to resource availability (review in Strier 2017). As the companions move across a landscape, small food patches may cause groups to break apart to forage (fission) to merge again later when going to sleep (fusion). While this pattern is relatively rare amongst primates, it occurs in hamadryas baboons, geladas and spider monkeys (Aureli et al. 2008).

Groups of the three members of the tribus Hominini – bonobos, chimpanzees, humans – likewise follow a fluid fission–fusion arrangement. However, unlike in hamadryas baboons, where splinter groups reunite before nightfall, all members of a chimpanzee group rarely, if ever, come together as a whole. Instead, the various foraging units spend the nights alone, spatially separated from others. Still, when smaller units meet, they may exchange members, so that, over a longer period of time, all group members get to see each other (Lehmann & Boesch 2003).

The standard model assumes that fission–fusion in chimpanzees is a means to mitigate food competition. Nevertheless, it has been suggested that communities split and then merge again for other reasons (reviews in Newton-Fisher et al. 2000, Anderson et al. 2002). These factors include (i) the overall abundance of food (e.g. Basabose 2004), (ii) the spatial distribution of food in patches (e.g. Chapman et al. 1995), (iii) the number of fertile females sporting ano-genital swellings (e.g. Goodall 1986, Stanford et al. 1994, Furuichi 2009), and (iv) the seasonal formation of hunting teams that prey on monkeys (Stanford et al. 1994, Boesch 1996). In reality, it is difficult to disentangle such a diverse portfolio of potential causes – not least, because environmental conditions can differ between years and communities. That being so, one should be cautious about concluding what is 'typical' for a given site.

Other things being equal, if food competition is mild, chimpanzees are hypothesized to maintain more predictable sizes of subgroups and higher relative levels of gregariousness (Lehmann & Boesch 2003). They mostly consume ripe fruits, a preference that is also true for the Rubondo apes (Moscovice et al. 2007). However, in notable difference from other, more seasonal sites, Rubondo does not go through a marked period of fruit scarcity (Chapter 3). While tree fruit abundance is somewhat seasonal, reduced from Jan–Mar, liana fruits are consistently abundant on the island (cf. Figure 3.10). Moreover, given that the Rubondo apes live under low-population density (see below), we expect relatively small levels of variation in grouping levels.

In the following, we employ the term *community* for the entire chimpanzee group, and we label the splinter groups as *parties*. It is customary to denote diurnal units as 'foraging parties', while nocturnal units that spend the night in sleeping platforms in close proximity to each other are termed 'sleeping parties' or 'nest groups'. Here, we label diurnal and nocturnal subunits in a neutral way as *day parties* and *night parties* – because these subgroups do not always forage or sleep, but may also rest and socialise, and to avoid semantic confusion with 'nests' of insects.

Michael Huffman – on Rubondo 2000–07

Figure B4.1 Michael A. Huffman on Rubondo. (Photo: Michael A. Huffman)

It had been my dream to work with wild chimpanzees since I was a little kid. One evening my mother was reading to me the tales of the monkey Curious George and I said to her that one day I was going to Africa to live with chimpanzees. I had forgotten about those words until my mother reminded me as an adult, but indeed that dream never faded. My research over the decades has had a focus on host parasite ecology, zoonotic transmission, self-medication and ethnopharmacology, with a special interest in traditional medicines acquired from observations of wild animals. I started my career studying chimpanzees at Tanzania's Mahale National Park in 1985 continuing there until 2003. It was in 2000 that Markus Borner invited me to Rubondo to "check the place out and see if the chimpanzees could be habituated and studied". We flew out together in his little plane via the Serengeti. I was intrigued by the island and visited several times. Of course, it was the hard work of Liza Moscovice, Klára Petrželková and Mwanahamisi Issa Mapua that made the research there during my involvement a success. Rubondo is a true paradise, the last untouched island in Lake Victoria. The chimpanzees, perhaps grudgingly at first, became the pioneers of a grand outdoor laboratory experiment to test the 'nature versus nurture' paradigm. This group of castaways created their own unique 'Rubondo culture', adapting to distinct circumstances, while yet amazingly acting in so many ways like their wild born counterparts. This island and these apes still hold many secrets for future generations of researchers.

Methodology of Tracking

During the main period of research into the sociality of the Rubondo apes (Oct12–Mar14), the chimpanzees were tracked by a team of 2–8 people consisting of JNM, local field assistants, foreign scientists, volunteers and at times VS. Half of the assistants had been involved in previous research with Liza Moscovice and Klára Petrželková in the 2000s. These trackers thus knew the forest and paths extremely well and could predict where the animals were likely to be found.

Tracking focussed on a slice of 30 km² in the northern Kamea and Masakela regions (Figure 4.4; cf. also Figure 3.2), albeit the chimpanzees ranged much more widely (see below). While trails created by elephants and hippopotami facilitated movement in the forest, many places – including chimpanzee nesting sites – were difficult to access because of thick under-story vegetation. Therefore, trails were cut in the two main survey areas, forming a grid system, 48 km in length, and marked every 100–300 m with a plastic red tag. Utilizing the grid and combining the collection of socio-ecological data with habituation efforts, two to three teams of 1–2 people surveyed the forests from dawn until dusk, with a rota ensuring that track-ing was done almost every day per week. The teams communicated with each other using walkie-talkies. Trackers searched near places where chim-panzees had last been met or were known to have previously visited and

Figure 4.4 Two main study regions in the island's north, with trail grids for chimpanzee surveys and straight-line botanical transects (Chapter 3). (Map: JNM)

listened for their vocalisations or drumming. Trackers would pay attention for about an hour, and if nothing was heard, moved to another spot at least 1 km away. Upon hearing sounds, the location was approached following compass bearings – while also recording indirect signs (nests in trees or on the ground, broken vegetation, feeding signs) and collecting samples (dung, hairs, food wadges). The location of chimpanzee sightings, trees and nests was recorded with GPS units (Garmin GPSMAP 62st, GPSMAP 62, eTrex H), which generally had a low error reading even under canopy cover.

Once in proximity and to not startle the apes, the trackers uttered a grunting noise. Having found them, attempts were made to follow the animals as long as possible. Habituation necessitates making contact with the same individuals repeatedly and at slowly decreasing intervals. However, given the often-poor visibility of frequently less than 15 m and the tendency of chimpanzees to avoid humans, it was difficult to gain their confidence.

Tracking was conducted for a total of 547 individual days, typically from 07:00–16:00 from Apr–Aug12 and 06:00–18:00 from Oct12–Mar14. Events from when the first chimpanzee became visible until the last disappeared out of sight were rated as an *encounter*. A new encounter was logged when the previous observation had occurred at least 3 h prior or at least 5 km from the previous sighting. On most occasions, the team did not find the animals again on the same day.

The size of *day parties* was defined as the number of chimpanzees travelling together that were seen and counted. An attempt was made to ascertain the number of individuals, and their sex and age–class. However, it is likely that some party members were missed, which would mean that sizes were underestimated.

The size of *night parties* was defined via nest clusters, i.e. the number of neighbouring arboreal sleeping platforms. These nests were located opportunistically by walking along the trail grid or by listening on hill tops for late evening or early morning chimpanzee vocalisations. The outcome of such nest counts depends on the definitions used to define a cluster. On the one hand, it is necessary to minimise the possibility of erroneously allocating night nests built on different days to the same cluster – as this would lead, incorrectly, to larger night party sizes. We therefore distinguished between newly constructed nests (<24 h old, identified from freshly broken foliage and the presence of faeces or urine underneath) and older nests (>24 h up to <30 days old, where leaves typically still retain their green coloration). This cautionary approach reflects experiences of other researchers that the animals often range in the same areas, and even sleep in the same trees, and the fact that some nests can persist in the environment for more than 100 days (Tutin & Fernandez 1984, Fowler 2006, Stewart 2011). On the other hand, night party size is a function of the cut-off distance for nests counted towards the same cluster. Previous researchers used intervals from any other nest of 15 m (Baldwin et al. 1981) or 30 m (Koops et al. 2007). Our cut-off distance is 50 m, which predisposes nest groups to be slightly larger. Still, while the shortest point between two nests was 1 m, the largest gap was 56 m between the two most peripheral newly built nests. Biases are unavoidable, if only that nests built in tall trees (> 30 m) are difficult to spot.

Data relating to direct encounters can tell us something about the degree to which the Rubondo chimpanzees get accustomed to human observers. Those entrusted with monitoring the released animals in the 1960s were keen to see them avoid or at least not seek out human contact – to minimise the risk of being injured by the ex-captives (Chapter 2). However, somewhat ironically, success in de-habituating chimpanzees is the opposite of what subsequent researchers hoped to achieve, i.e. direct observations. In any case, the Rubondo apes became increasingly wary of humans, latest from the 1990s, with the result that researchers were hardly ever able to see them. For example, during 3 months of fieldwork on Rubondo in 1974, Gustl Anzenberger saw the animals only three times and for only a few minutes (Müller & Anzenberger 1995: 5), whereas they were seen only a single time by Guido Müller during his 2 months of research in 1994 (ibid., p. 53). In 2002–04, the monthly mean of encounters ranged from 2.4–5.9 (Moscovice 2006), while in 2006–07, rates hovered around two sightings per month (Klára Petrželková, pers. comm.). Nevertheless, when located, some groups were followed for up to 4 h at a distance of 30–50 m (Moscovice 2006).

Habituation is unattainable with such low encounter rates, as success depends upon the number of neutral contacts with each individual chimpanzee (Boesch-Achermann & Boesch 1994, Sommer et al. 2004). Our own attempts between Apr12 and Mar14 fared comparably better – but not so much better in that the apes did actually become accustomed to our presence.

Encounter Rates

The team observed chimpanzees most often in 2013 in Apr–May and Jul–Sep, but also during all other months except Sep12, albeit with great variation. Tracking teams had mostly 2–4 members, but at times included 7 or 8 people. Interestingly, the number and duration of encounters per month increased when teams were larger. This may be related to more trackers having 'more eyes'. However, in addition, chimpanzees might have been extra suspicious if they felt that a single person tried to sneak up on them. A larger team might feel less threatening, perhaps because these people somehow seem to relate to each other (cf. Sommer et al. 2004).

With, on average, 0.2 observations per day, an encounter happened every fifth day (monthly mean = 7, median 4, sd = 7, range 1–22, n = 146). The length of staying in contact ranged from < 1–695 min (mean = 45 min, median 15 min, sd = 84, n = 146). The maximum time of almost 12 hours refers to Nov12 when the research team had contact with two animals for an entire day, as the apes never left a tree and remained in one spot. Over the course of the study, the rates of sightings per month increased markedly, but the duration of the encounters increased less. This result, disappointing in terms of habituation success, might simply reflect that the attempt was too short.

At other sites, it took 6–8 years to get apes used to human observers (Goodall 1986, Boesch & Boesch-Achermann 2000). Unfortunately, the main sponsors of our fieldwork did not subscribe to such a long-term perspective, and discontinued their funding with respect to habituation. Following on from our own fieldwork, TANAPA has made continued efforts to habituate the chimpanzees, employing experienced trackers such as Simon Nkuzi and James Leonard. While encounter rates have improved, it is still not possible to follow the apes from nest to nest consistently on a daily basis (James Leonard, pers. comm. to JNM).

Party Sizes

As for the sizes of subgroups (Figure 4.5), day parties typically encompassed four individuals (mean=4.4, median=4.0, sd=3.3, range 1–26, n=146 encounters). Most sightings discovered a single individual, followed by 3, 5 and 8 animals (Figure 4.5 a). As for night parties, 1,224 nests in various stages of decay were detected – with the distribution of cluster size categories similar to those of day parties (Figure 4.5 b). Of these, 190 sleeping platforms

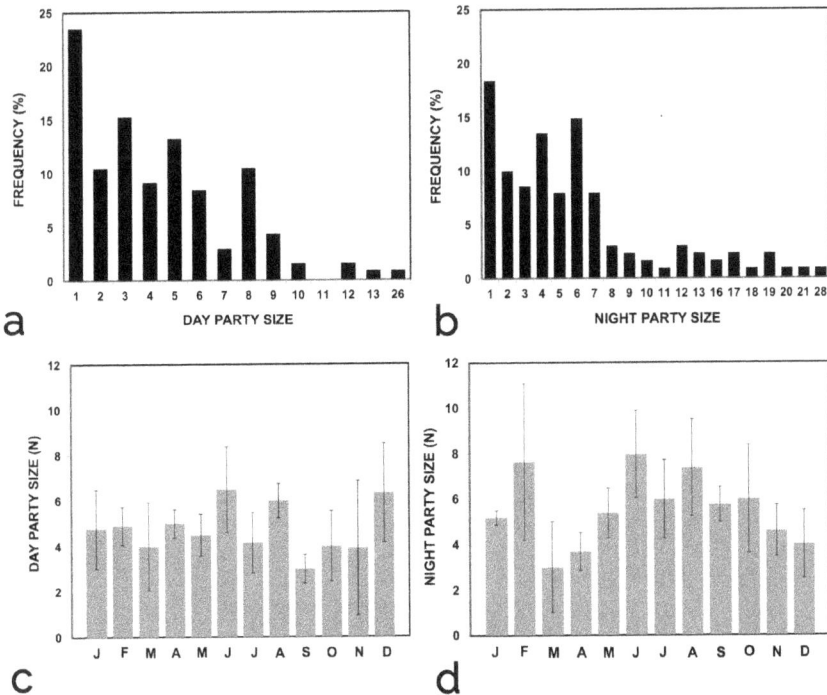

Figure 4.5 Chimpanzee party sizes (Apr12–Mar14). (a) Day party frequencies. (b) Night party frequencies. (c) Annual distribution of day parties. (d) Annual distribution of night parties.

(16%) had been constructed the previous evening. Thus, for night parties, we have two measures – one that counts all nests in what seems like a cluster (a method prone to lump nests built during different days), and less error-prone counts of fresh nests only (found with faeces or urine underneath). The inclusive counts of all nests produced figures somewhat larger than day parties (mean 5.8, median 5.0, sd = 5.0, range 1–28, n = 143 clusters).

The comparably bigger night aggregations could indicate that independently travelling foraging parties sometimes unite in the evening. This mechanism might be facilitated by calls, which are particularly frequent in late afternoons, presumably allowing scattered foraging units to connect. Calling peaks also early in the morning, when chimpanzees descend from their night trees. Alternatively, smaller day parties might be an artefact of not registering all apes that travel together in such parties, because of obstruction by the vegetation or because the animals actively avoid being seen. Another bias could be the inflation of nests built during different evenings into the same cluster-count. The explanation of smaller day parties joining for the night is the more likely one, as counting fresh nests only led to night parties of typically five individuals (mean 5.4, median 4.0, sd = 5, n = 38 clusters), i.e. a figure very similar to the all-nest counts. In fact, there was no significant difference in night party sizes based on nests 1–30 days old or fresh nests <24 h (Mann-Whitney-Wilcoxon rank sum test with continuity correction, $W = 1982$, $p = 0.626$). Nevertheless, while night parties were somewhat larger than day parties, this difference was, overall, not significant (Spearman, rho = -0.065, s = 1215, $p = 0.7894$).

Liza Moscovice – on Rubondo 2002–04

Figure B4.2 Liza Moscovice (top l.) with her team of trackers, including Joseph Mgwesa (bottom r.). (Photo: Michael Huffman)

While on a study abroad program in Tanzania in 1996, during a visit to Kigoma, I fortuitously met Michael Huffman, famous for his research on chimpanzees and ethnobotany. Chatting on the shores of Lake Tanganyika, he told me about his idea to study the little-known chimpanzees of Rubondo. Six years after that first encounter, I began my doctoral studies on the island. One of my goals was to identify habitat features and chimpanzee behaviours that had allowed the descendants of captive animals to create a life for themselves, in a novel environment, without the influence of wild conspecifics. I also planned to study chimpanzee sociality and material culture but had to re-align my expectations half-way through the project, because it was very difficult to find the animals in those 240 km² of forest and to observe them directly. The strongest memory to this day is still that feeling of sitting on the ground in the forest, trying to be as quiet as possible, and feeling the intense and curious gaze of a chimpanzee watching me.

Measures of gregariousness were previously reported for 1994 (Müller & Anzenberger 1995: 22f, 51) and for 2003 (Moscovice et al. 2007). Together with our 2012–14 research, this generates three data sets spanning 20 years, separated by roughly a decade each:

- 1990s: day parties (no data), night parties (mean 3.8, range 1–24, 392 nests of any age in 102 clusters; mean 5.5, range 2–8, 22 fresh nests in 4 clusters);
- 2000s: day parties (mean 3.3, median 3, range 1–9, 56 sightings), night parties (mean 3.4, median 2, range 1–19, 138 clusters of fresh nests);
- 2010s: day parties (mean 4.4, median 4.0, range 1–26, 146 sightings), night parties (mean 5.8, median 5.0, range 1–28, 143 clusters of fresh nests).

Given the likely growth of the population over the 20 years covered by these data, we might have expected party sizes to increase. However, while the figures are broadly similar, no increase was registered. Instead, the means of clusters of fresh nests fluctuate from 6 to 3 to 6, which might reflect stochasticity due to different sampling periods or different methodologies.

In any case, Rubondo party sizes of 3–6 are broadly similar to means reported for native populations, e.g. 4–5: Gashaka, Nigeria; 4: Bossou, Guinea; 5: Kanyawara, Uganda; 6: Sonso, Uganda; 6: Gombe, Tanzania; 6: Mahale, Tanzania (Boesch & Boesch-Achermann 2000, Jesus 2020).

Seasonal Effects

Party sizes during our 2012–14 research did not significantly fluctuate across the year (Kruskal-Wallis median-test, day parties: $\chi^2 = 20$, df $= 20$, P $= 0.458$; night parties: $\chi^2 = 18$, df $= 18$, p $= 0.456$) (Figure 4.5 c, d). Our findings are in agreement with those for 2003, as party sizes did also not change across months (Kruskal-Wallis, $\chi^2 = 17.40$–20.05, df = 18, p = 0.118–0.496; Moscovice et al. 2007).

While neither day nor night party sizes varied dramatically between months, we nevertheless compared them to ecological factors, i.e. rainfall, humidity, temperature, and fruit availability (Table 4.1). For 90% of the measurements, the environment had no discernible effect. The only significant association was a negative correlation between rainfall and night party sizes. Perhaps, rain disrupts the tendency of foraging parties to join up for the night; the chimpanzees may instead refrain from the extra travel necessary to meet their companions, nesting on their own close to their last foraging patch.

The salient feature emerging from our analyses is the fact that the gregariousness of the Rubondo chimpanzees is scarcely influenced by weather or food availability. These findings reflect that Rubondo's climate is, by and large, non-seasonal (Chapter 3), at least compared to other sites.

For example, at Gashaka in Nigeria, roughly half of the year has very little to no rain, while the other half sees heavy downpours. Here, while party size fluctuation was measured with the same method as on Rubondo, night parties become significantly bigger when fruit availability increases, while day party sizes remain constant throughout the year (Jesus 2020: 139). This result might seem surprising, because one might expect that more animals will forage together when more fruit are available. However, the disparity of constant day party sizes and fluctuating night party sizes is likely caused by foraging constraints. Hence, food patches rarely occur in large clusters with many trees and vines fruiting in close proximity. Instead, fruit-bearing plants are spaced out, represented by single trees or vines, and only seldom in groves of interconnected plants (Brockman & Schaik 2005). As a consequence, while food competition is generally reduced in times of

Table 4.1 Variation in day and night party sizes as a function of environmental variables (Apr12–Mar14)

Environmental variable (x)	Party size variable (y)	R value	Spearman test	P value
Rainfall (mm)	Night (n)	0.609	−0.819	0.001
	Day (n)	0.002	−0.224	0.507
Temp min (°C)	Night (n)	0.024	−0.440	0.152
	Day (n)	0.082	−0.024	0.945
Temp max (°C)	Night (n)	0.021	0.239	0.453
	Day (n)	0.005	0.199	0.558
Humidity min (%)	Night (n)	0.314	−0.573	0.051
	Day (n)	0.028	−0.094	0.784
Fruit index [a]	Night (n)	0.197	−0.537	0.072
	Day (n)	0.023	0.103	0.764

[a] Sum of DBH of transect fruiting plants, i.e. trees and lianas (for which the host tree's DBH was used); cf. Chapter 3

plenty, this only holds true for the habitat as a whole, but not for a particular food patch. Chimpanzees are therefore inclined, even when the number of specific fruiting trees increase, to continue to forage in small parties (similar to observations at Budongo, Uganda; Newton-Fisher et al. 2000: 625). In contrast, larger night parties do form at Gashaka during peaks of fruit availability – because the more food patches exist, the shorter are the distances between each patch, allowing various parties to forage in proximity. Therefore, given reduced travel costs, day parties can merge more easily when nightfall is approaching, as opposed to periods when patches are more spaced out. Compared to these constraints recorded at Gashaka, Nigeria, the situation is fundamentally different on Rubondo, because food availability hardly fluctuates; as a consequence, day parties and night parties maintain relatively similar sizes throughout the year.

Still, as mentioned earlier, chimpanzee fission–fusion dynamics are likely influenced by a variety of factors, from weather to the availability of plant food or animal prey and periodicity of sexual activity. Such multi-factorial effects may explain why, at some sites, and in contrast to the arguments we just provided, day parties do seemingly fluctuate with habitat-wide fruit production (Gombe, Tanzania: Wrangham 1977, Goodall 1986; Mahale, Tanzania: Matsumoto-Oda et al. 1998; Kibale, Uganda: Chapman et al. 1995, Mitani et al. 2002; Taï, Ivory Coast: Boesch 1996).

The take-home message from our analyses of the basic pattern of gregariousness in the Rubondo chimpanzees is that the dimensions and fluctuations of their night and day parties are within the range of variation reported elsewhere. With this, these introduced apes have retained – resp. 'reinvented' – the typical fission–fusion association patterns of endemic populations.

Head Counts: Populating Paradise

While the Rubondo chimpanzees occasionally tolerate the presence of human observers, they are not individually known. We can therefore not infer the number and age–sex composition of the current island residents by listing its members – as is possible at sites where the animals are fully habituated. Instead, to understand how the population developed since the initial releases, we need to resort to indirect means.

In principle, we could apply genetic profiling based on microsatellites, i.e. tracts of repetitive DNA, which, because of their high mutation rates, allow individual identification. The DNA can be extracted from the same biological samples, such as faeces, which we used to determine the geographical origin of the founder animals (Chapter 2). However, employing microsatellites is expensive, and a lack of funds meant that this approach has not yet been realised.

An alternative technique is the mathematical assessment of population growth. Such simulations require quantitative input about the life–history traits of the organisms in question. In both captive and wild settings, chimpanzees are roughly similar in terms of late maturation, long inter-birth intervals (IBIs), extended investment in offspring and slow population growth rates, with females giving birth to roughly five infants in their lifetime (e.g. Boesch & Boesch-Achermann 2000: 64, Sugiyama 2004, Littleton 2005). Thus, the basic reproductive parameters (see literature in Behringer et al. 2014) are as follows, with figures indicating mean and range (r) of days (d) or years (y): (i) *menstrual cycle*: 32–36 d, menstrual flow typically visible; around ovulation, females develop pinkish ano-genital swellings; (ii) *pregnancy*: 228 d (r 202–261 d); single infants, rarely twinning; (iii) *age of weaning*: 4 y; some infants may seek their mother's proximity until 10 y; (iv) *birth interval*: 4–6 y (r 2–8 y); (v) *age of first birth*: 13 y (r 11–20 y; r 5–32 y in captivity); (vi) *male sexual maturity*: 13–14 y; (vii) *mortality 1st year of life*: 22% for males, 18% for females; (viii) *life expectancy*: in captivity, about 32 y for males and 38 y for females; while lifespans are shorter in the wild, males can survive till their mid 50s, and some females above 60 y (Wood et al. 2017).

Computer programmes employ parameters about a starting population to arrive at a projection for some point in the future. The R package 'mmage' simulates dynamics of animal populations using a discrete-time matrix model (Lesnoff 2015), while 'VORTEX' (Lacy 1993) simulates the viability of a particular population. Both programmes consider stochastic factors such as individual demographic characteristics (lifespan, sex), and variation in population-level birth and death rates. In addition, VORTEX can include deterministic factors such as predation pressure, disease, weather, or hunting frequency. However, for Rubondo, main environmental factors are stable, even across decades; for example, the chimpanzees are not affected by large predators, conspecific competition, habitat destruction or hunting.

Still, a serious problem with these simulations is their susceptibility to small variations in life-history parameters, which, given that projections cover several decades, can produce widely differing results (Msindai 2018). For example, one set of models was fed with a value of a relatively low fertility rate (15%) and high mortality rates, which led to a projected decline of numbers from 14 founders in 1969 to around 6 (model VORTEX) or 13 individuals (model mmage) in the year 2013. Another model fed with a somewhat higher fertility rate (25%), arrived at forecasts of 11–28 individuals in the year 2019. In yet another set of simulations, mortality rates for communities in Kibale, Uganda (Kanyawara, Muller & Wrangham 2014; Ngogo, Wood et al. 2017) were plugged into the model, which, for the year 2019, projected numbers between 26 and 73. These fluctuating results highlight the problem of projecting future numbers based on values obtained from other populations instead of the one which is being modelled.

Not using mathematical simulations, Monica Borner (1985) assumed that the island harboured "at least 20" chimpanzees in the mid 1980s. Müller & Anzenberger (1995: 57–65), without employing computer programmes, but considering reproductive parameters from wild communities elsewhere, projected 32 (min 24, max 40) animals for the year 1997. Huffman et al. (2008), for the early 2000s, conjectured a figure of "35+" individuals.

Given the difficulties with simulations and the absence of direct counts of identified or genotyped individuals, we resorted to another approach to estimate population growth. For this, we rely on the maximum numbers of apes sighted together, respectively the largest number of night nests found in a single cluster – as these counts represent reliable figures about the minimum population size.

The data point for direct sightings relates to 24Jul13, when trackers observed two separate day parties. One team (consisting of visiting researcher Paco Bertolani and several field assistants) surveyed the central Nyakutukula hills. Here, in grassland with few trees, i.e. an area with good visibility, they came across 26 chimpanzees. The encounter began with 3 apes, but gradually more and more of them aggregated – most likely because some of them were consuming an infant sitatunga (Chapter 5). Flimsy day-nests on the ground were also found. The team stayed with the apes for several hours, video-recording parts of the event. While the sex of most animals remained unknown, there were at least 6 adult females and 4 adult males, plus a number of juveniles. On the same day, JNM, together with a field assistant, surveyed the island's far northern region of Kasenya, where they observed 1 adult male and 1 adult female with an oestrous swelling. They stayed with the apes for about 30 min, but heard calls for several hours afterwards. The two separate encounters took place 12 aerial km apart from each other. From these direct sightings, we can ascertain that the island harbours a minimum of 28 chimpanzees, including at least 7 adult females and 5 adult males.

The largest single nest cluster was recorded on 09May13 by JNM and several field assistants in the island's northern region of Kamea, consisting of 28 nests. These were likely built during the preceding days because the assemblage had not been present during intensive surveys of that forest section a week before. The cluster consisted of 26 tree nests and 2 elaborate nests on the ground, which also seemed to have been used during the night. Infants or small juveniles do not construct their own nests and instead sleep with their mothers. The proportion of infants in well-studied communities across Africa averages 23.6% (minimum 11.2%, maximum 38.1%; Hughes et al. 2011: Table 14.3; see Chapter 6). We thus need to add 24% to the largest night nest count, which leads to the conclusion that the island harbours a minimum of $(28 + 6.7 =)$ 35 chimpanzees.

Our maximum counts are larger than values for the year 2003, where the most numerous day party contained 9 animals and the biggest night cluster

19 nests (Moscovice et al. 2007). Still, our counts based on a direct sighting (28 animals) resp. night nest cluster (35 animals) might be underestimations, given the reclusive and shy nature of mother–infant pairs, who tend to travel apart from larger gatherings of conspecifics (Wrangham & Smuts 1980, Goodall 1986).

If we nevertheless apply our count of 35 animals, then a starting number of 14 founders in 1969 had grown by 21 individuals over the next 46 years, which equates to an increase of 150% or 3.3% per year. This situation compares favourably with other long-term study sites, where death rates exceed reproductive rates, causing communities to decline. For example, the mortality over the next 10 years of all infants born is 79% in Taï, Ivory Coast and 55% in Gombe, Tanzania (Hill et al. 2001). The few communities known to have increased in size in the past few decades include Kanyawara, Uganda, with 10-year mortality reaching only 53% for females and 67% for males (Muller & Wrangham 2014), and Nogogo, Uganda, with 33% for females and 19% for males (Wood et al. 2017). We can suspect that Rubondo's chimpanzees benefit from similarly higher probabilities of survivorship.

The founder pool contained more females (n = 9) than males (n = 7; of which at least 2 were shot dead). We have reliable information about the proportion of males and females 50 years on, given that the sex was determined for 195 biological samples (Chapter 2), and that we can assume that our collection procedure was random. Of these samples, 59.5% derived from females and 40.5% from males, which translates into proportions of 21 females to 14 males for the 2014 population. This inferred male/female sex ratio of 60%/40% is almost identical with endemic communities, given that males consistently experience higher mortality, causing a preponderance of females (e.g. 59.5%, Taï-East, Ivory Coast, year 2014/15, McCarthy et al. 2018; 56.4%, Ngogo, Uganda, year 2016, Wood et al. 2017; 62.5%, Sonso, Uganda, year 2001, Reynolds 2005).

A combination of advantageous conditions has allowed the Rubondo apes to multiply (see discussion in Chapter 6; cf. also Huffman et al. 2008). To begin with, the surviving founder animals benefitted from the absence of human activities such as poaching, pastoralism, agriculture, horticulturalism and tree-cutting as well as zero levels of human–wildlife conflict exacerbated by crop-raiding (Arcus Foundation 2015, 2018). Unlike what happened on neighbouring islands, Rubondo's forest cover has not been degraded. On the contrary, the island's status as a national park has provided effective protection against destructive anthropogenic actions (Mwambola et al. 2016). The relative geographical isolation on an island probably also shielded the Rubondo apes from outbreaks of communicable diseases transmitted by wild or domestic animals or humans. Moreover, given the absence of large terrestrial carnivores, the islanders have little to fear from predators, notwithstanding some danger, particularly for infants and juveniles, from martial eagle and crowned eagle, Nile crocodile and up to 5 m long pythons (Kade 1967, Borner 1985; JNM pers. obs.). Importantly, the Rubondo founders were not confronted with native conspecifics. As we will discuss (Chapter 6), given the xenophobic nature of chimpanzees, numerous individuals released at other sites have been killed by resident apes.

Furthermore, elsewhere, dispersing females tend to delay reproduction, due to stress when entering a new community (Sugiyama 2004, Walker et al. 2018). As we will argue in the subsequent section, the Rubondo apes have not split into different communities, and females thus remain and breed in their familiar social unit. Additionally, in the absence of inter-group competition, they might experience shorter inter-birth intervals and higher rates of offspring survival (Lemoine et al. 2020). Moreover, given evidence from captivity (Ely et al. 2005), hybrid vigour might have suppressed mortality, at least in early generations albeit the relatively small numbers of founder females combined with intra-community reproduction could in the future lead to inbreeding depression (Huffman et al. 2008).

On Rubondo, a mean party size of 4.4 in a community of 35 members equates to a share of 13%, again well in the range of relative party sizes for other sites (e.g. 12%: Sonso, Uganda; 13%: Taï, Ivory Coast; 11–16%: Gashaka, Nigeria; Boesch & Boesch-Achermann 2000, Sommer et al. 2004: 312). However, we know that mean party sizes become relatively smaller when the community is bigger, perhaps because there is more feeding competition overall (Boesch & Boesch-Achermann 2000). Thus, the big community at Ngogo, Uganda, maintains average party sizes of 6.7, which is large in absolute terms, but small in relative terms, translating to just 4.8% (Wakefield 2008). The Rubondo figure of 13% therefore characterises the community as a still relatively small unit.

In sum, the surviving members of the cohorts released onto Rubondo have found themselves in rather 'paradisiacal' conditions that allowed for high rates of population growth.

Ranging Far and Wide: Not a Must, But a Can

Our research also shed light on the question, how much of the island the Rubondo chimpanzees are utilizing, and, whether or not, half a century after the founder animals were introduced, their descendants constitute a single or several communities.

Territoriality

Chimpanzee groups at forested sites occupy much less terrain than those in arid open habitats, where trees and fruits are sparse. Population densities per km^2 vary accordingly, reaching up to 7 individuals in some forests (Yamagiwa 1999), whilst less than 1 individual in more open habitats

(Baldwin et al. 1982). But despite tree-rich habitats containing more food and thus necessitating less travel, ranging in forests can also be constrained because of competition with conspecifics, or, increasingly frequent, with humans. As a result, when multiple communities reside in close proximity, home-ranges overlap by 30–50% (Herbinger et al. 2001).

In general, chimpanzee females, once reaching sexual maturity with an average 12 years, tend to transfer into a new group, presumably to avoid inbreeding. However, in fragmented habitats where the next group lives at a far distance, females may stay in their natal groups (Goodall 1986). For example, in Gombe, Tanzania – a national park that is an enclave within a human-shaped landscape – only between 13% and 50% of females were found to transfer out (Goodall 1986, Pusey et al. 1997). Such exceptions notwithstanding, relationships amongst females are weak or antagonistic, because as emigrants from different natal communities, they lack close ties of kinship.

This contrasts with males, who are philopatric, meaning they remain in their natal group. Males therefore represent a 'band of brothers' (or half-brothers or cousins), although these blood relations do not prevent them from competing over access to mates. The main expression of kin-mediated cooperation amongst males comes in the form of rivalry with neighbouring groups. These contests can be extremely violent, including the killing – and cannibalistic consumption – of overpowered victims, events that have been labelled as 'lethal raiding' and 'proto-warfare' (e.g. Goodall 1986, Wrangham & Peterson 1996, Mitani et al. 2010). This deadly violence is thought to reflect adaptive strategies, in that killers gain fitness benefits by increasing their access to food and females. A recent compilation for 18 communities with 152 witnessed, inferred or suspected mortal encounters supported this explanation (Wilson et al. 2014). Most killings (66%) involved intercommunity attacks, males were the most fre-quent attackers (92% of participants) and victims (73%), and attackers greatly outnumbered their victims, with a median 8:1 ratio. Because of such inter-community bellicosity and because male chimpanzees patrol the border of their range, it is appropriate to apply the term 'territory' with respect to the terrain they occupy.

The threat of conspecific aggression may lead to temporary avoidance of fruit-rich feeding areas when neighbours happen to be there (Nishida 1968). This causes a more intensive utilization of smaller core areas within the larger territories (Herbinger et al. 2001, Williams et al. 2002, Newton-Fisher 2003). Thus, at most chimpanzee study sites, inter-community dynamics influence the overall extent as well as the patterns of habitat use.

Rubondo island, roughly 240 km² of which 80% is forested, with little human disturbance and no native apes present, therefore makes for an inter-esting place to examine how chimpanzees utilize space when 'traditional' constraints do not apply.

Measuring Range Use

Varied approaches aiming to quantify the extent of primate home-ranges may yield very different results (e.g. Lehmann & Boesch 2003). Methodologically, the basic problems centre around the decision whether or not to include landscape parts that are avoided or not used by the animals (lacunas), such as water bodies (rivers, small lakes, swamps) or barren zones (cliffs, rocky outcrops). Additional concerns rest with 'transit' areas (corridors), and what to do with rarely visited, remote positions (outposts). Decisions for or against the inclusion of such spaces will yield home-range sizes which do not differ by percentage points, but orders of magnitude. We have to keep this general issue in mind, because it renders cross-site comparisons difficult.

Much of our detailed socioecological research focussed on the Masakela and Kamea regions in the island's north. The decision to concentrate on these segments was not meant to imply that the chimpanzees did not utilize other slices of the island. In fact, two previous studies had established that they were also active elsewhere. On this account, during 1994 and the 2002–08 study periods, while nests were also found in Masakela and Kamea, they were likewise detected far north towards Kasenya and in the central sectors of Irumo and Nyakutukula (Müller & Anzenberger 1995: 47, Moscovice et al. 2010). Being aware of this, over the 23 months from Apr12 to Feb14, in addition to our studies at Kamea and Masakela, we conducted regular opportunistic visits to other parts, noting whether or not chimpanzees had utilized them. In particular, our surveys were extended beyond the designated core-research area, when, after several months of almost daily sightings of chimpanzees in Kamea and Masakela, encounter rates fell dramatically in Jun13. In addition, hardly any newly constructed night nests were detected – leading us to conclude that most apes had gone elsewhere. Consequently, the trackers conducted intense surveys to find out where the apes had moved to – first in north-western Kasenya, but then also further south. In fact, a month or so later, in Jul13, the trackers encountered a large party in the south, confirming findings of previous researchers about a larger extent of the range.

The total range-related records amounted to 146 direct observations, 93 times when vocalisations were heard, 143 nest clusters (representing 1,224 night nests in various stages of decay, from freshly constructed to old), 22 ground nests, plus 76 additional incidences, when faeces, hair or food wadges were found (Figure 4.6).

Our accumulated data about localities frequented by the chimpanzees informed three approaches to estimate home-range sizes:

• The Kernel Density Estimation (KDE; Worton 1989) plots the data and creates a curve of the distribution, calculated by weighing the distance of all the points in each specific location along the distribution.

- The Local Convex Hull (LoCoH; Getz & Wilmers 2004) estimates not only the size of the range, but also constructs a utilization distribution, representing the probability of finding the animals within a given area. This method is better at detecting 'holes' not used by the animals and allowing their exclusion from the area calculation.
- The Minimum Convex Polygon (MCP; Jennrich & Turner 1969) is a simpler method. It can be visualised as a board with a set of nails on it – indicating localities where evidence for the animals has been recorded. When a rubber band is stretched around all the outermost nails, the resulting shape represents the convex hull. By selecting levels from 100% down, one can remove extreme outliers or more peripheral points. Still, the rubber band might not touch all of the nails, thus neglecting the internal structure of the range. The MCP can be further manually modified to exclude land locations where the study animals were not recorded. The restricted polygon then becomes concave.

The different methods can produce markedly different results for range size and core areas of utilization. In addition, KDE and MCP methods will include high type II error rates and overestimate the range when unvisited areas are not excluded. As for our Rubondo calculations, the unused areas of the surrounding waters of Lake Victoria were manually removed, while unvisited areas on the island itself were not excluded – because, as the saying goes, 'absence of evidence is not evidence of absence'. By elaborating upon these issues, we draw attention to the problematic nature of inter-site comparisons, given the extreme sensitivity to methodology.

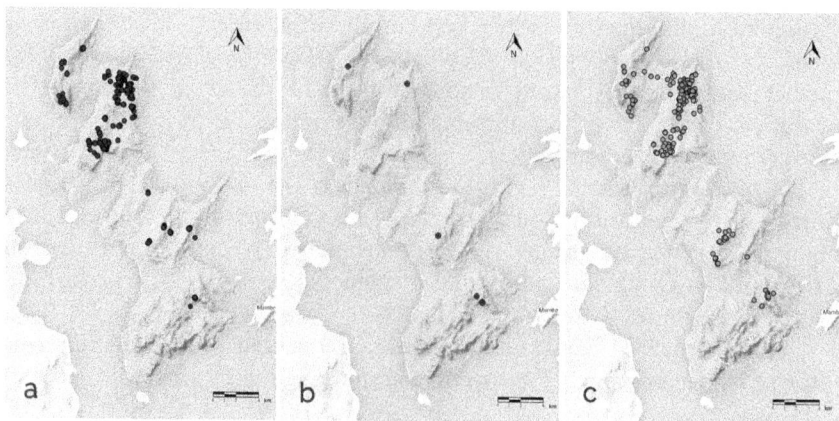

Figure 4.6 Locations of evidence for chimpanzee presence (Apr12–Mar14). (a) Tree night nests (n = 1,224). (b) Ground night nests (n = 22). (c) Daytime sightings (n = 146). Because of close spatial association, evidence marker dots will often overlap.

Extent of the Rubondo Range

As a case in point, the varied approaches for Rubondo yielded very different results:

- KDE: level 50% = 19.0 km², 95% = 115.0 km²;
- k LoCoH: 41.1 km²;
- MCP: level 50% = 12.4 km², 95% = 70.0 km², 99% = 95.1 km² (Figure 4.7).

Thus, based on selected parameters, the returned values for range use vary considerably. The sizes of core areas fluctuate from 12.4 km² (50% MCP) via 19.0 km² (50% KDE) to 41.1 km² (k LoCoH) – with the last figure representing the most appropriate measure. As for the overall range, the returned figures vary between 95.1 km² (99% MCP) and 115.0 km² (95% KDE).

However, while the KDE figure translates into an overall use of 49% of the island's 237 km², this is still an underestimate, as a comparison

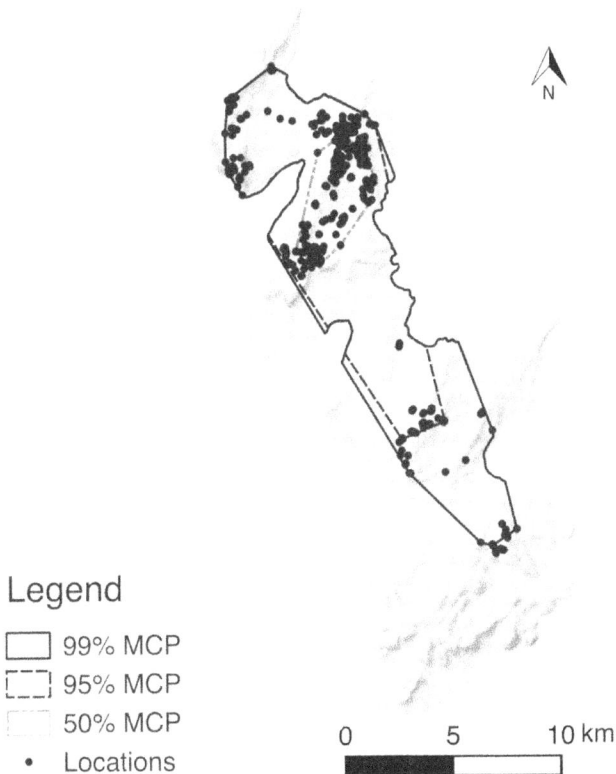

Legend
- 99% MCP
- 95% MCP
- 50% MCP
- • Locations

0   5   10 km

Figure 4.7 Home-ranges calculated via a minimum convex polygon (MCP), adjusted to exclude Lake Victoria. Dots indicate location of chimpanzee evidence (Apr12–Mar14).

with other study periods suggests. While no range estimation was provided in the 1994 survey (Müller & Anzenberger 1995), the 2003 study utilized 455 data points of chimpanzee evidence to arrive at 99% MCP values (Moscovice et al. 2010). Accordingly, considering only the survey areas (Kasenya, Masakela and Kamea, Nyakutukula) and the straight-line distance required to travel across them, the range size was determined as 43 km². The maximum range estimate, including the survey areas and all intervening habitat, came to 82 km². During the 2003 study, rangers reported a chimpanzee sighting far outside of the main tracking regions, near a southern ranger post (probably Lukaya), suggesting that the animal made a brief excursion. A 100% MCP estimate that includes this sighting leads to a maximum ranging area of 119 km². Consequently, range sizes deduced from the 2003 study vary from 43 to 119 km² – a factor of 3 in magnitude.

Still, there can be little doubt that calculated ranges reflect biases in terms of survey efforts – or of survey periods. Thus, we visited areas further south than Nyakutukula not previously covered by other researchers, in hills closer to the Lukaya post. Here, indeed, we found additional evidence for chimpanzee presence (cf. Figure 4.6, Figure 4.7). Moreover, a single animal was also seen further south, in the Lukukuru ranger post area, sometime before 2012 (TANAPA rangers, pers. comm. to JNM), and the 1994 surveys found nests in yet other regions not identified as chimpanzee ranges in 2012–14 or 2003. Therefore, if we combine the recorded evidence for the 1990s, 2000s and 2010s, we can ascertain that the chimpanzees, at one or another time, criss-cross the whole island from the far north to the far south (Figure 4.8).

Preferred Areas

This synopsis also reveals distinct overlap, demonstrating that, over two decades, the Rubondo chimpanzees preferred the same island parts. The favourite segments were separated by corridors 5–9 km apart (Kasenya to Masakela 5 km, Masakela to Kamea 6 km, Kamea to Nykutukulu 9 km, Nyakutukula to Lukukuru 6 km), with a 24 km straight-line distance between two most northern and southern areas. That discovery raises the question as to what may explain the establishment of such 'traditional' extents of ranging.

For sure, during all study periods, certain island parts were avoided as nest sites, namely, the lowest altitudes at the coasts, the northern grasslands and lowland mixed forests in the central and southern regions. Conversely, preferred were the more northern, central and southern *Diospyros* forests that grow on hills. The only hilly part lacking chimpanzee evidence were the most southern ridges towards the Lukukuru ranger post. Thus, the animals clearly prefer forested segments which tend to be found at higher elevations – an expected result, as fruit from trees and associated lianas constitutes

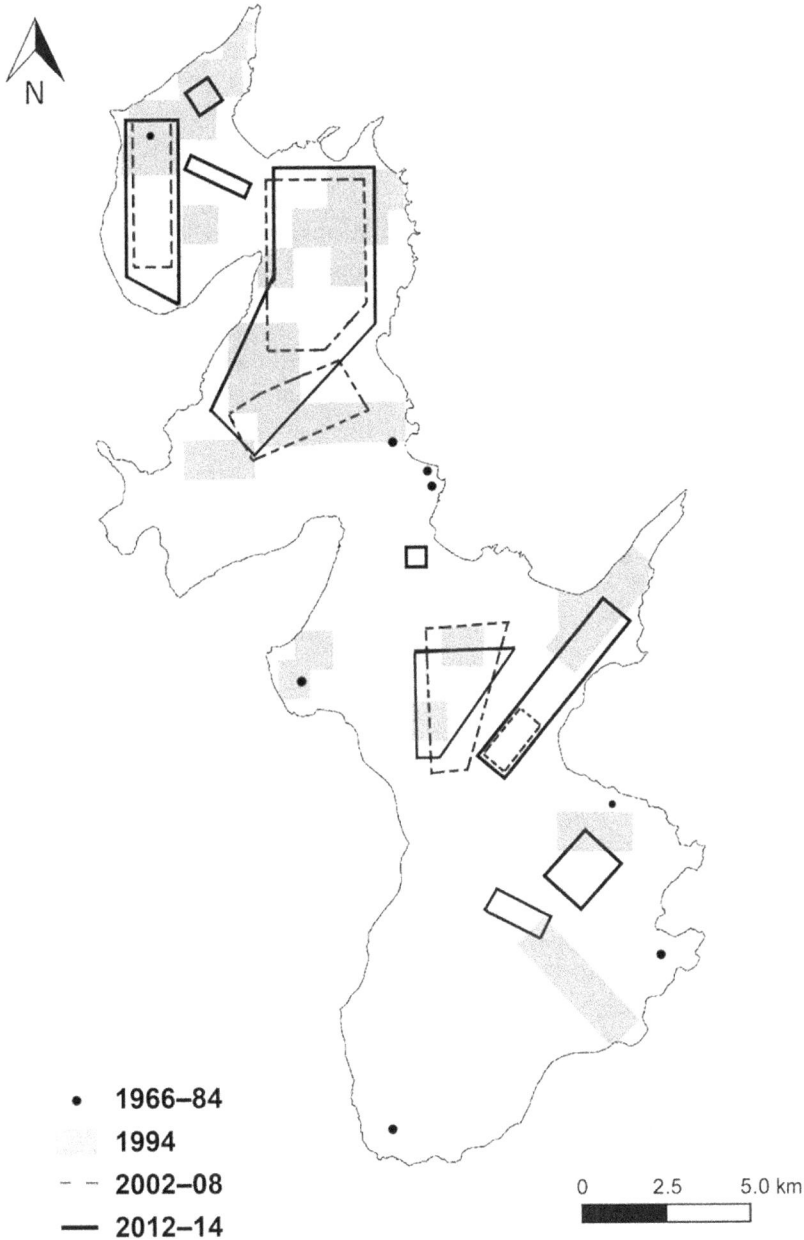

Figure 4.8 Main areas of chimpanzee activity as discerned during different study periods: 1966–84 (occasional sightings); 1994 (approximate location for n = 392 nests, Müller & Anzenberger 1995: 23, 56); 2002–08 (n = 317 GPS points, L. Moscovice and K. Petrželkova, pers. comm.; 2003, Liza Moscovice et al. 2010); 2012–14 (n = 1,183 GPS points, Msindai 2018).

the main food at all study sites, with tree canopies also providing the best opportunities to construct sleeping platforms (Chapter 5).

At some endemic sites, ranging patterns reflect area-distinct availability of plant fruit (e.g. Chapman et al. 1997, Furuichi et al. 2001). Because of the temporally uneven tracking of the habitat, our own study did not explore such potential seasonality. However, the 2003 study compared plant composition and productivity across forest blocks separated by several kilometres (Moscovice et al. 2010: 2–17; Chapter 5). It was hypothesized that chimpanzees would move towards the south during Oct–Dec to feed on the then abundant and preferred *Drypetes gerrardii* fruit, while residing in the north from Mar–May when two other preferred fruit species occur there at high densities. However, there were no seasonal differences in the daily utilization of sectors, as quantified by mean proportions of chimpanzee encounters in each part during: first dry season (42.8%, sd 7.3, χ^2 (2, n = 43) = 0.634, p = 0.728); first wet season (42.8%, sd 2.4, χ^2 (2, n = 95) = 0.192, p = 0.909); long dry season (33.8%, sd 8.8, χ^2 (2, n = 114) = 2.063, p = 0.356) and second wet season (38.2%, sd 12.0, χ^2 (2, n = 98) = 3.140, p = 0.208). This suggests that regionally distinct and seasonal fruit availability had little impact on habitat utilization. These findings contrast with those from endemic sites, where, as a likely strategy to reduce feeding competition, smaller subgroups are formed when fruit is scarce (Wrangham 2000). However, party sizes of Rubondo chimpanzees show no seasonal fluctuation (cf. Figure 4.5), suggesting that food competition is considerably less severe.

Still, some seasonal preference might exist, for example based on our own experience in Jun13 that virtually all chimpanzees left the northern regions (i.e. Masakela and Kamea). Elephants forage in most, if not all island parts with high chimpanzee activity (JNM, pers. obs.; Moscovice 2006: 142, Mwambola et al. 2016). Inter-specific interactions could influence chimpanzee ranging, directly – because elephants are food competitors – and indirectly – because the apes might avoid encounters with the pachyderms.

Within a given sector, there is little evidence that some localities are being favoured as sleeping sites and others to forage – as has been observed elsewhere (van Lawick-Goodall 1968, Ghighlieri 1984; cf. suggestions in Müller & Anzenberger 1995: 47). For example, when chimpanzees visit the edges of their range to feed, this might entail an increased probability of disturbance by humans or neighbouring communities, leading the apes to retreat to core areas to spend the night there. However, the Rubondo residents used the same parts of their range to forage and to build tree nests or ground nests (cf. Figure 4.6).

A Single Community at Low Population Density

Given this relaxed regime, a follow-on question relates to whether or not the Rubondo apes form a single community or multiple. The 1994 study suggested that there were 4–5 groups, given the spatial separation of pre-ferred areas. These 'Gruppen' were believed to reflect family or maternal ties (Müller & Anzenberger 1995: 52). However, this is a rather unortho-dox interpretation of the multimale–multifemale societal structure of chim-panzees. Similarly, the 2003 study raised the possibility of the existence of "three distinct and non-interacting social groups and/or solitary individuals on the island", because of ranging habits of three identifiable males. These adults were seen in all preferred areas, but, despite such overlap, were never recorded there on the same day or in the same party, "suggesting some form of temporal avoidance" (Moscovice 2006: 131). These inferences come with the caveat that the observations do not allow a conclusive con-firmation. In fact, the authors later reconsider, stating that they "strongly suspect that the entire chimpanzee population utilized all three of the survey areas" (Moscovice et al. 2010: 2717). While we do not have reliable data on individual identities, our observation in Jun–Jul13 of many animals moving from the north to the south, even if anecdotal, supports the view that the population constitutes one single community. This is also borne out by the fact that there has never been any evidence for major aggressive interactions nor indications of males 'patrolling' their range as they do elsewhere. There is simply no discernible border, as everybody seems to belong to the same community.

All these findings, from stable party sizes to non-seasonal habitat usage to the absence of competing conspecifics tie in with the large dimension of the Rubondo ape range. Hence, even the most conservative estimates of 43–119 km^2 (Moscovice et al. 2010) resp. 95–115 km^2 (our study) – far exceed reported values for native populations at forested sites elsewhere (Table 4.2). Combining our lowest range estimates with the projected num-ber of chimpanzees yields estimates of a population density of less than 1 animal/km^2. In fact, apart from a brief lack of preferred food, the animals face virtually no constraint should they wish to visit any part of the island. With this, a community of 35 animals on an island 237 km^2 would translate into a population density of 0.1 animals/km^2. Such figure sets the Rubondo apes clearly apart from other forested sites.

We thus agree with Moscovice et al. (2010), that, in the absence of strong seasonal plant fruit productivity and competition from other apes or humans, space utilization by the Rubondo inhabitants is based almost entirely on habitat favoured for foraging and nesting. In other words: the Rubondo apes do not move to certain areas, because they must. They criss-cross the island widely because they can.

Therefore, even in the absence of detailed behavioural observations we can conclude that the fission–fusion and sex ratio dynamics as well as the ranging pattern of Rubondo animals fall into the normal range of variation. Despite the fact that the founder individuals only as infants experienced life as a 'wild' chimpanzee, they and/or their descendants developed a type of sociality very similar to endemic communities. It would be interesting to

Table 4.2 Home-range sizes, community members and population density at selected chimpanzee study sites

Country	Community	Study period	Range (km²) (a)	Community members (n)	Density (animals / km²)	Reference
Uganda	Budongo	1994–95	7	42	6.0	Newton-Fisher 2003
Uganda	Ngogo	2003–06	28	143	5.1	Amsler 2009
Tanzania	Gombe	1975–92	11	51	4.6	Williams et al. 2004
Ivory Coast	Taï South	1996–97	27	63	2.3	Herbinger et al. 2001
Ivory Coast	Taï North	1996–97	17	35	2.1	Herbinger et al. 2001
Uganda	Kanyawara	2007–09	27	48	1.8	Bertolani 2013
Nigeria	Gashaka	2012	26	35	1.3	Sommer et al. 2004
Uganda	Kanyawara	1996–98	38	50	1.3	Wilson et al. 2007
Ivory Coast	Taï Middle	1996–97	12	11	0.9	Herbinger et al. 2001
Tanzania	Rubondo	2003	43	35+	0.8	Moscovice et al. 2010
Tanzania	Rubondo	2012–14	95	35	0.4	Msindai 2018

(a) Most ranges measured via a minimum convex polygon (MCP)

know to what degree this development toward chimpanzee-typical pattern is brought about by inherited factors or environmental stimuli.

In spite of everything, the Rubondo apes, without being able to emulate or copy wild comrades, and with no human support, were successful in rehabilitating themselves.

5 Embedded: Mastering a New Environment

Figure 5.1 Nest built by a Rubondo chimpanzee, roughly one month old. (Photo: JNM)

Once released into the unknown, the founders of the Rubondo population had to figure out on their own how to sleep in a tree and what foods to eat. This was far from trivial, as even such fundamental behaviours require considerable skill and knowledge, information that is normally learnt through experience and passed down the generations, constituting a specific local 'primate culture'. In addition, an animal's physiology is normally well adapted to a routine nutritional input, rendering a shift in diet challenging.

In this chapter, we explore which nest building skills, dietary habits and foraging strategies the Rubondo apes ultimately developed, how their

DOI: 10.4324/9780367822781-5

gastrointestinal system coped with parasites, and we describe the structure of their gut microbiome. Thus, how have the Rubondo apes managed to be integrated members of an island ecosystem – a mosaic of adaptations that includes 'bed-making' for the night (Figure 5.1), and other day-to-day functionalities required to be truly 'embedded' in the environment?

Nesting Habits: Securing a Good Night's Sleep

Many animals, from insects to birds to mammals, create shelters – burrows, hives, dens, nests – to exert some control over their environment for the benefit of themselves and their offspring (overview in Hansell 2007). Whether shelters are tools (Fruth & Hohmann 1996) or even "super tools" (Matsuzawa 1991), is a matter of scientific debate, given that the final object is not "directly manipulated in its entirety" once completed (Shumaker et al. 2011). In any case, the creators of such artefacts are "ecosystem engineers" (Jones et al. 1994), given that they impact not only upon themselves, but upon their surroundings and with this on other species as well.

In the primate order, shelter construction has evolved 6–8 times, particularly in lemurs and galagos (Kappeler 1998), while it is absent in tarsiers, new world monkeys and old world monkeys – albeit some taxa such as callitrichids and tarsiers make use of existing cavities, e.g. tree holes. Given that all extant great apes (chimpanzees, bonobos, gorillas, orangutans) construct temporary arboreal structures from bent branches termed *nests* (Nissen 1931), this form of elementary technology likely evolved after the line leading to the Hylobatidae separated from the Hominidae. However, whilst other primate nest builders occupy them for extended periods, to sleep in and raise young, great apes generally use the same nest only once. Therefore, it has been suggested that these constructions should be referred to as *beds* (Hiraiwa-Hasegawa 1989) or *sleeping platforms* (McGrew 1992).

Great apes, each evening around dusk as well as sometimes during the day, build a new nest to sleep in or rest (Baldwin et al. 1981, Kano 1983, Tutin & Fernandez 1984, Fruth & Hohmann 1993, Fruth 1995, Fowler 2006, Hernandez-Aguilar 2006, Koops 2011, Stewart 2011). After climbing a suitable tree, the apes will identify one or two strong limbs as foundation, onto which they break and bend several branches, interlocking them and then folding smaller twigs over the edge to form a rim; the finished structure is a circular bowl, and sometimes leafy twigs are placed inside as a lining. Typically, a nest takes around 3–8 min to construct. Occasionally, apes will build less elaborate nests on the ground, often only raking together loose vegetation, to serve as daytime resting spots.

Given a lifespan that can exceed 50 years, a chimpanzee may build more than 20,000 nests, spending about 12 h/day in these structures. First attempts occur at 8–12 months of age but young apes do not make their own nests until after weaning at around 3–5 years. Captives are often deprived of opportunities to build nests. However, if provided with suitable material, they tend to build them, suggesting an inborn component to the behaviour (Videan 2006, Locke et al. 2011).

Although the principle of construction is uniform, techniques vary across and within populations, reflecting flexible responses to seasonality and available material (Baldwin et al. 1981). Moreover, chimpanzees are selective in the tree species they use, the position at which nests are placed in the tree and their location in the landscape. It is not clear if these differences reflect efforts to mitigate particular environmental constraints, although nest building has been associated with a variety of potential functions. These fall into half a dozen categories, related to comfort, thermoregulation, disease control and predator avoidance, or just being stopovers near food sources or information hubs (e.g. Baldwin et al. 1981, Fruth & Hohmann 1996, Stewart 2011).

At the minimum, compared to crouching in a naked tree fork, sleeping inside a nest is probably more secure as it guards against accidental falls. In general, given their large body size, nest building may permit apes to rest in tree locations that would otherwise not support their weight. This includes constructions close to terminal branches, which might be difficult to access for leopards. In addition, nests can buffer against cold temperatures or wind, since a bowl-shaped structure reduces the amount of body surface exposed to the air. The cushioned configuration is also more comfortable, enabling a good sleep, important for restorative processes that benefit brain and body. Moreover, sap exuding from broken branches or leaves might have properties that repel insects, including parasite-transmitting varieties.

Some other potential functions would not require the actual construction of sleeping platforms, but are served by them nevertheless. Thus, as new nests are built each day, this prevents the accumulation of parasites. Also, nests are perhaps built on slopes where there is lower insect density due to orographic winds. Apes may also prefer to sleep close to food sources to reduce travel costs – albeit they should avoid resting in trees that actually contain fruit, because other nocturnal foragers would disturb them at. Finally, nest clusters enable apes to spend the night in close proximity, exchanging information about, for example, their physical conditions.

Data Sets

We explore the nest building behaviour of the Rubondo apes and compare it with native populations elsewhere to assess some of the functional explanations. For this, we recorded locations, nesting tree species and features of architecture, whenever possible integrating our own data with those of previous researchers (Chapter 4). The total sample comprises three sets, roughly separated by 10-year intervals, with a total of 2,145 nests:

- 1994: 392 nests (2 months of fieldwork, Müller & Anzenberger 1995: 24, 30);

- 2004–07: 529 nests (13 months of fieldwork, Petrželková et al. 2010a);
- 2012–14: 1,224 nests (22 months of fieldwork, Msindai 2018).

Due to differing methodologies, sample sizes vary for some variables (e.g. nest height, nest tree species). Comparability between the data sets is also compromised because they cover island sectors of distinct vegetation structure (Chapter 3), and entries are often not controlled for seasonality. Moreover, nest heights during the 1994 and 2004–07 surveys were estimated by eye-sight. In addition, the 1994 data were collected by rangers with varying field skills, and nests higher up were found difficult to detect. Our 2012–14 study, on the other hand, worked with field assistants of considerable expertise, and vertical measurements obtained via a Haglöf electronic clinometer.

Nest Heights

The vast majority of recorded nests were built in trees (97.2%), while ground nests amounted to 2.8%, with some variation between study periods (0.0% in 1994; 7.2% in 2004–07, n = 38; 1.7% in 2012–14, n = 21) (Figure 5.2). It is likely that the majority of ground nests were built during

Figure 5.2 Nest types on Rubondo. (a) One of the rare ground nests (2013). (b) Three arboreal nests (indicated by arrows) built in the same tree. (Photos: JNM)

Figure 5.2 Continued

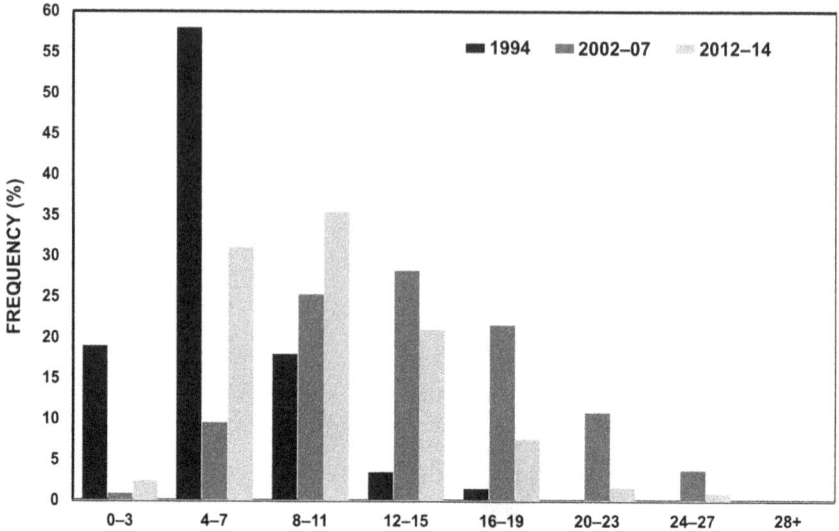

Figure 5.3 Height distribution classes of arboreal nests during three study periods (1994, Müller & Anzenberger 1995: 25, n = 389; 2002–07, n = 240, Klara Petrželková, pers. comm.; see also Petrželková et al. 2010a; 2012–14, Msindai 2018; this study n = 1,173). Category breakdown was determined by the 1994 data (0–3 = 0–3.99 m, 4–7 = 4.0–7.99 m, 8–11 = 8.0–11.99 m; etc.).

the day, as short-lived resting spots, assembled from low branches, shrubs, grasses and saplings.

Across the study periods average nest heights vary between 7.0 m (1994), 14.0 m (2003–07) and 11.2 m (2012–14). A breakdown of categories like-wise reveals considerable differences (Figure 5.3), with a strong preponder-ance of lower-height nests in 1994, while the 2004–07 and 2012–14 findings are in greater agreement. The discrepancies are likely due to the mentioned methodological differences. Still, while the absolute values across studies differ according to method, the general trend of values found for certain regions of the island is similar (Table 5.1). Hence, in the central sectors, dur-ing independent study periods, nests were found to have been built on aver-age, 3.4 m higher up in the trees than in the northern and far-northern island parts. It is likely that this finding reflects different forest physiognomies.

We restrict comparisons with heights measured at other study sites to the 2012–14 data as these were enabled by a clinometer. This synopsis reveals that the Rubondo nests are built at the lowest heights, compared to other chimpanzee habitats:

- median 10 m (mean 10 m, range 1.8–30.8 m, Rubondo, Tanzania, Msindai 2018, this study);

- median 10 m (mean 12 m, range 2–45 m, Lope, Gabon, Wrogemann 1992);
- median 11 m (mean 11 m, range 0–16 m, Mt. Assirik, Senegal, Baldwin et al. 1981);
- median 11 m (mean 11 m, range 0–35 m, Seringbara, Guinea, Koops et al. 2012);
- median 12 m (mean 13 m, range 5–35 m, Sapo, Liberia, Anderson et al. 1983);
- median 20 m (mean 13 m, range 5–35 m, Taï, Ivory Coast, Fruth 1995);
- median 21 m (mean 22 m, range 3–43 m, Conkouati, Rep. of Congo, Farmer 2002a).

Specific causes for the low nest heights on Rubondo are difficult to ascertain. They may be related to vegetation structure (e.g. canopy tops on Rubondo might be less suitable to construct nests in) or to the absence of large terrestrial predators – which may likewise explain why the general sleeping locale is frequently the same (more than 50%, Petржelková et al. 2010a).

The variation of nest height on Rubondo could be explained 70% of the time by tree height, given that nests higher up were located in taller trees (Figure 5.4). A good predictor for tree height is DHB (diameter at breast height), because thicker trees tend to be taller (Figure 5.5). This points towards a causal connection between trunk diameter and the probability that a tree will be selected as a resting spot – suggesting that stability is an issue. The assumption is supported by various lines of evidence. To begin with, arboreal nests overwhelmingly relied on tree parts, and only occasionally integrated or primarily used lianas. The vast majority – 86.3% – of arboreal nests were restricted to a single tree, while nests integrating two trees accounted for 12.5% and those with three trees for only 1.1%. Interestingly, the DBH of trees for non-integrated nests (mean 33 cm, sd 22, n = 923) was significantly larger than the stem diameter in case of integrated nests (2 trees 27 cm, sd 22, n = 126; 3 trees 19 cm, sd 20, n = 12) (Kruskal-Wallis: $\chi^2 = 33$, df = 4,

Table 5.1 Sector-specific height of arboreal nests built by Rubondo chimpanzees during three study periods (cf. Figure 5.3)

Height (m)	Whole island	Far north, north			Central		
	1994	2003–07	2012–14	Mean	2003–07	2012–14	Mean
Median		11.8	9.5	10.7	16.0	12.1	14.1
Mean	7.0	11.8	10.1	11.0	16.1	12.3	14.2
Min	1.5	3.5	2.0	2.8	6.0	3.2	4.6
Max	20.0	25.0	30.6	27.8	25.0	24.0	24.5
n	389	103	1098		121	72	

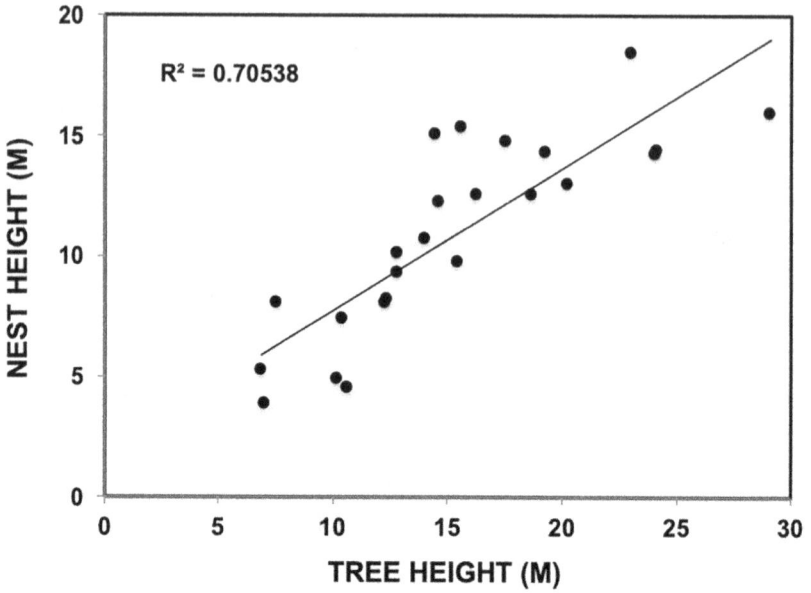

Figure 5.4 Night nest height as a function of tree height (n = 690).

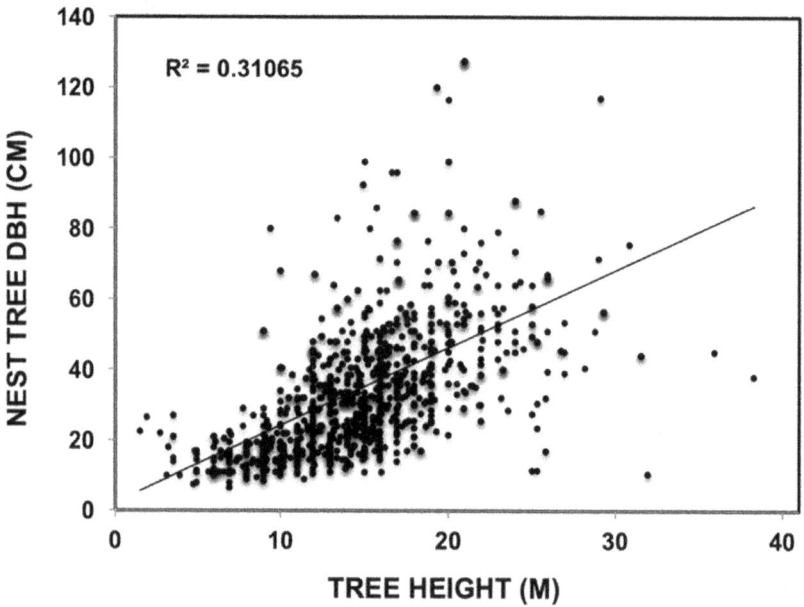

Figure 5.5 Nest tree DBH (diameter at breast height) as a function of tree height (n = 803).

n = 997, p < 0.001). It thus seems likely that thinner trees are less suitable to support a chimpanzee's body weight, which then requires combining several to gain the same support as from one larger and thicker trunk. In general, nest tree DBH (mean 34 cm, median = 30 cm, max = 222 cm, sd = 21, n = 996) was significantly larger than the DBH of 'random' transects trees (mean 23 cm, median = 16 cm, max = 236 cm, sd = 18, n = 1160) (Wilcoxon test with continuity correction: 95% CI: 143 to 148, df = 147, p < 0.001). A preference to nest in trees with a larger trunk girth than the average of all available trees was also found at other sites (Fruth 1995, Fowler 2006b, Stewart 2011, Koops 2011).

Nesting Tree Species

The Rubondo chimpanzees utilized 31 tree species to nest in (Table 5.2), with an overlap of 56.3% between the 2004–07 and 2012–14 study periods. Most data for the first period were collected in the centre of the island, where the preferred species is *Drypetes gerrardii* (40.6%), while latter period was biased towards the north, were *Synsepalum brevipes* was preferred (63.9%). We do not know what spurs the animals to seek out these species, i.e. whether their branches are particularly easy to bend, if their leaves make for good cushions, or if they are 'spacious' and allow the construction of several nests in one canopy.

While there is great variation across sites, the Rubondo findings are above the reported minimum of 23 nest tree species (Lagoas de Cufada, Guinea-Bissau, Carvalho et al. 2015) but much lower than the reported maximum of 115 species (Seringbara, Guinea, Koops 2011). Generally, there is little overlap in nest tree species across Africa, which includes the use of introduced taxa such as eucalyptus (*Eucalyptus grandis*), guava (*Psidium guajava*), cocoa (*Theobroma cacao*) and Caribbean pine (*Pinus caribaea*) (McCarthy et al. 2017). This points to flexibility in the construction ability of chimpanzees, or, perhaps more parsimoniously, that many taxa are suitable to build nests in – an effect that may be aided by the considerable diversity of trees on the African continent (Chapter 3).

Nevertheless, chimpanzees appear to be highly selective in terms of nesting trees. At Rubondo, just three species accounted for 82.5% of all nest building (*S. brevipes* 50.6%, *D. gerrardi* 23.5%, *Pancovia turbinata*, 8.5%). This preference is time-resistant, as the same species topped the lists during different study periods separated by about 10 years. A preponderance of more than 80% is far stronger than what has been reported elsewhere where the top three species make up a considerable share of nesting trees (59%, Bwindi, Uganda, Stanford & O'Malley 2008; 52%, Gashaka, Nigeria, Fowler 2006; 50%, Kahuzi-Biega, DR Congo, Basabose & Yamagiwa 2002; 46%, Issa, Tanzania, Stewart 2011).

On Rubondo, the proportion of *P. turbinata* is similar to its share on the random line transects (17.2%). However, the transect shares of the two

Table 5.2 Nesting tree species (n = 31) identified during the study periods of 2004–07 (n = 25) and 2012–14 (n = 24)

Family	Latin name	Common name	2004–07 (n)	2004–07 (%)	2012–14 (n)	2012–14 (%)	Average %
Sapotaceae	Synsepalum brevipes		169	37.3	821	63.9	50.6
Putranjivaceae	Drypetes gerrardii	Bastard white ironwood	184	40.6	81	6.3	23.5
Sapindaceae	Pancovia turbinata		22	4.9	155	12.1	8.5
Lecaniodiscus	Lecaniodiscus fraxinifolius	River-litchi	15	3.3	66	5.1	4.2
Dichapetalaceae	Dichapetalum stuhlmannii	Goat-killer	2	0.4	27	2.1	1.3
Rutaceae	Teclea nobilis		2	0.4	20	1.6	1.0
Capparaceae	Maerua duschesnei	Common bush-cherry	3	0.7	17	1.3	1.0
Ebenaceae	Diospyros abyssinica		16	3.5		0.0	1.8
Sapotaceae	Mimusops kummel	Red milkwood	5	1.1	10	0.8	0.9
Verbenaceae	Premna angolensis		2	0.4	12	0.9	0.7
Sapindaceae	Haplocoelum foliolosum	Northern galla-plum	2	0.4	9	0.7	0.6
Verbenaceae	Vitex doniana	Black plum	7	1.5	4	0.3	0.9
Euphorbiaceae	Alchornea birtella	Christmas bush		0.0	10	0.8	0.4
Rubiaceae	Canthium lactescens	Hairy turkey berry	2	0.4	7	0.5	0.5
Rubiaceae	Craterispermum schweinfurthii		2	0.4	7	0.5	0.5
Rubiaceae	Rothmannia urcelliformis			0.0	9	0.7	0.4
Guttiferae	Garcinia huillensis		1	0.2	7	0.5	0.4
Moraceae	Antiaris toxicaria	Bark cloth tree, Antiaris, sacking tree	1	0.2	5	0.4	0.3
Moraceae	Morus mesozygia	Black mulberry	1	0.2	5	0.4	0.3
Moraceae	Ficus sansibarica	Fig tree	3	0.7	2	0.2	0.4

Family	Species	Common name	n	%	n	%	%
Rhamnaceae	*Maesopsis eminii*	Musizi, umbrella tree	5	1.1		0.0	0.6
Anacardiaceae	*Pseudospondias microcarpa*	African grape	1	0.2	4	0.3	0.3
Euphorbiaceae	*Alchornea laxiflora*		2	0.4		0.0	0.2
Polygalaceae	*Carpolobia conradsiana*		2	0.4		0.0	0.2
Ulmaceae	*Celtis africana*	White stinkwood		0.0	2	0.2	0.1
Euphorbiaceae	*Croton sylvaticus*	Forest fever-berry		0.0	2	0.2	0.1
Ebenaceae	*Diospyros mespiliformis*	African ebony, jackal-berry		0.0	2	0.2	0.1
Sapindaceae	*Zanha golungensis*		2	0.4		0.0	0.2
Mimosoideae	*Albizia gummifera*	Peacock flower	1	0.2		0.0	0.1
Flacourtiaceae	*Flacourtia indica*	Batoka plum		0.0	1	0.1	0.0
Bignoniaceae	*Markhamia* sp.		1	0.2		0.0	0.1
Sum			453	100.0	1285	100.0	100.00

Note: 'n' refers to the number of nests built in specimens of a particular species

TREE SPECIMENS (%)

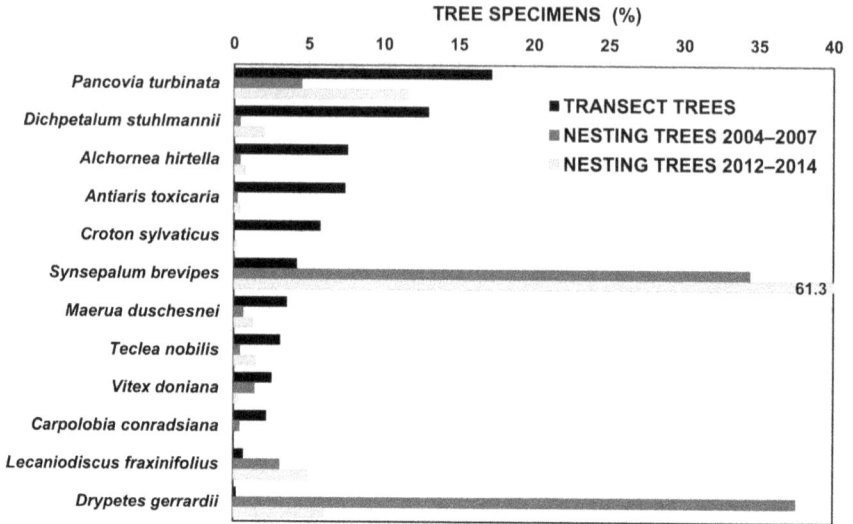

Figure 5.6 Shares of preferred species of nesting trees compared against their occurrence on random transects (Chapter 3).

other top species are much lower (*S. brevipes* 4.2%, *D. gerrardii* 0.2%), indicating a strong selectivity in favour of these taxa (Figure 5.6).

The 50.6% preference for *S. brevipes* almost matches the strongest proportion for a single-species reported so far, i.e. the 53% with which *Dialium guineense* was the preferred tree at Lagoas de Cufada, Guinea-Bissau (Carvalho et al. 2015). Elsewhere, shares of species preferred on Rubondo are very different. Thus, at Nimba, Guinea, *Synsepalum* sp. and *Drypetes* sp. were used only 3–5% of the time (Koops 2011), while at Issa, Tanzania, *D. gerrardii* was used 2% of the time (Stewart 2011). Conversely, at Kibira, Burundi, *Parinari* species accounted for 15% of all nests (Hakizimana et al. 2015), whilst Rubondo chimpanzees avoided these trees altogether. Thus, we are still far away from understanding whether or not there is a cross-continent preference to nest in certain tree species.

Climate and Physical Environment

At some sites, nests are built higher up in the canopy during rainy days (Assirik, Senegal, Baldwin et al. 1981; Fongoli, Senegal, Pruetz et al. 2008; Lope, Gabon, Wrogemann 1992). This has been related to the benefit of drying off quicker once rain has stopped, as less water drips from vegetation overhead and the rising sun will heat up such tree portions earlier in the day. However, no such correlation exists on Rubondo. Indeed, the mean height of nests varies significantly across months from 8 m (Jul) to 11 m (May) (Kruskal-Wallis test: $\chi^2 = 192$, df = 157, p = 0.0299), but this is not connected to a significant correlation of monthly nest height with rainfall (Pearson's correlation: r = –0.01, df = 531, 95% CI –0.095 to 0.075, p > 0.8), and neither with temperature or humidity, or across seasons (wet vs. dry). Still, in months with higher rainfall, nests were built in shorter trees (Pearson's correlation: r = –0.106, df = 521, 95% CI –0.190 to –0.021, p = 0.015), similar to a 1.5 m difference recorded at Nimba, Guinea (Koops 2011).

In addition, chimpanzees at some sites build more open nests in more rainy periods (Baldwin et al. 1981, Wrogemann 1992, Stewart 2011). This has likewise been linked to reduce exposure to water dripping off from vegetation above once rain has stopped. However, empirical evidence for such connection is lacking (Koops 2011). In fact, on Rubondo, the opposite is true, as in months with higher rainfall, nests were not only built lower in the canopy, but were also more often covered with vegetation overhead. In addition, Rubondo residents were nesting in taller trees and higher up in the canopy during colder and more humid nights – a measure certainly not suited to increase ambient temperatures within the nest.

Still, there might be a causal connection between weather conditions and the altitude at which nests are built (Figure 5.7). The shoreline on Rubondo is at 1,100 m a.s.l., and hills rise to 1,486 m. On average, nests were built at

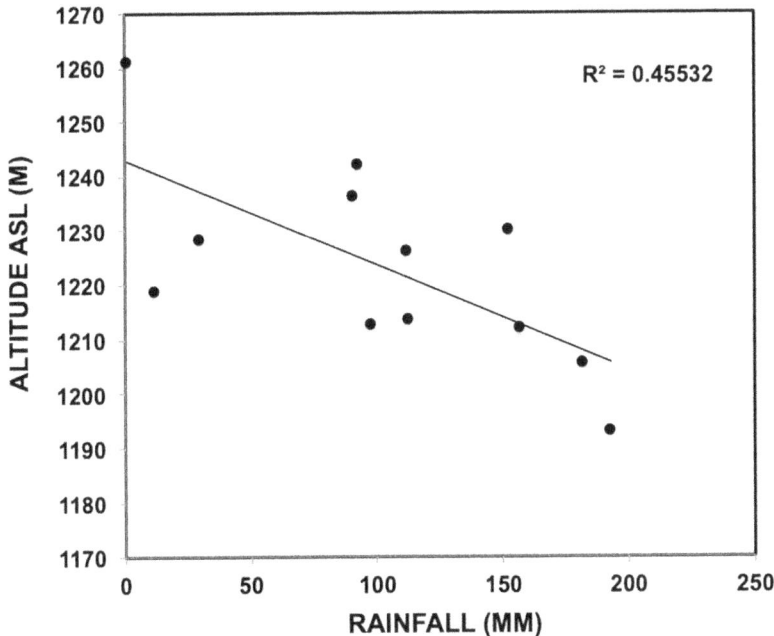

Figure 5.7 Altitude at which nests were built (n = 801) as a function of monthly rainfall.

1,228 m a.s.l. (median = 1,228 m, sd = 2, n = 801), with three quarters (75%) being found at 1,200–1,259 m a.s.l.. However, altitude differed significantly between study months (Kruskal-Wallis: $\chi^2 = 198$, df = 103, p < 0.001), with 1,260 m during a particularly dry month (Jul), and 1,190–1,220 m during wet months (Mar, Dec). Hence, the Rubondo chimpanzees are perhaps choosing to nest at a lower altitude in order to stay warmer during wet nights. However, at Nimba, Guinea, they tend to nest at higher altitudes in the wet season (Koops 2011).

These disparate associations between climatic factors and nest traits render it difficult to postulate unambiguous causal connections.

Potential Functions of Nest Building

What, if anything, can our findings add to the question whether architectural nest traits mitigate particular environmental problems?

Some features can be accounted for by simple ecological corollaries without underlying functional aspects, such as, for example, that nest height is lower in shorter trees. On the other hand, chimpanzees over-select trees that are thicker – and thus taller. This not only offers more opportunities to find a suitable spot in the canopy, but also one that provides a more solid foundation for the structure – compared to thinner and less-sturdy specimens.

Despite the fact that large terrestrial predators are absent on Rubondo, the apes still nest above ground – in 97% of all cases. The 3% of structures on the ground may all be day nests (Petrželková et al. 2010a). That the average arboreal nest was 10 m high might indicate efforts to minimise disturbance from roving elephants, which can reach a shoulder height of 3 m and are often active in parts of the island likewise frequented by the apes (Mwambola et al. 2016).

In terms of association with thermoregulation, our results are likewise ambiguous. We found no evidence that, during rainy periods, chimpanzees aim to reduce water dripping from overhead, by building nests high up in the canopy and open to the sky. In fact, the opposite is true, as during rains, nests heights were lower and more often covered by vegetation overhead.

Preferred nesting trees may also provide preferred food (see below). At Nimba, Guinea, this overlap is 32% (Koops 2011). We might expect that chimpanzees are not sleeping in those trees when they are actually fruiting – to avoid disturbance by nocturnal foragers such as bats or squirrels. Yet, on Rubondo, the apes did nest in particular trees when fruit was ripening (Petrželková et al. 2010a), albeit we have not quantified the overlap.

There is also no clear-cut connection between disease-mitigation and nest building. In general, there might be fewer mosquitoes 10 m above ground – which could explain the preference for arboreal sleeping sites (Koops 2011). Moreover, previously built nest are very rarely re-used. On Rubondo, of 529 nests recorded in 2004–07, this was recorded only twice (0.4%, Petrželková et al. 2010a) – indicating a potential effort to avoid accumulation of disease vectors. In addition, green plant material contains volatile secondary compounds with biocidal effects on parasites and pathogens (Dubiec et al. 2013). Orangutans, during periods of high mosquito density, prefer nest trees with known mosquito-repellent properties (Largo et al. 2009). Might this also guide the behaviour of Rubondo's chimpanzees?

The original human inhabitants of the island used bark and leaves of *S. brevipes to* fend off malaria (Kiwango et al. 2005), while elsewhere, fevers are treated this way (Odugbemi 2008). However, *S. brevipes* is not a preferred nesting tree at other sites. Moreover, for only a fraction of tropical plants – perhaps 2% – are their actual chemical properties known. This, together with the fact that local people ascribe a multitude of health benefits to tropical plants, renders it difficult to link the preference for specific nesting trees to medicinal properties (Koutsioni & Sommer 2011).

Quite independent from any discussion about the functions of nests themselves, nesting tree species and nesting in general, the introducees to Rubondo island, and in particular their descendants, developed a pattern of nest building that is well within the variation of native populations.

Diet Diversity: A Palate for Plants

Most chimpanzees dwell in rainforests, albeit some persist in rather arid conditions – but all of them are predominantly herbivores. Their habitats are not homogeneous and even adjacent fragments can show marked disparities in fruit production (Janmaat et al. 2016). For example, in Kibale, Uganda, the Ngogo community range covers more resource rich patches than that of the neighbouring Kanyawara community, allowing the former to maintain a three times larger population density (Watts et al. 2012). There may also be marked intra- or supra-annual fluctuations in food availability, plus variations brought about by asynchronous or synchronised plant reproduction.

Within these constraints, chimpanzees consume about 45 food items at arid sites but more than 300 in forests (review in Harrison & Marshall 2011). There is general agreement that the *preferred food* is ripe fruit (taking up 56–71% of foraging time) while *filler fallback food* is represented by leaves (18–21%) as well as bark, pith, flowers, gum and terrestrial herbs (11–23%). Figs are the major *staple fallback food* (up to 91%), although this characterisation is a matter of debate, because figs can also constitute a preferred food (Dutton et al. 2014).

In addition, chimpanzees consume some animal matter. At some sites, vertebrates are preyed upon (0.3–6% of feeding time), comprising at least 25 species (80% colobus monkeys, 20% mammals such as duikers, bush pigs, baboons and rodents). Occasional attacks on human infants by wild chimpanzees have also been registered, with victims being eaten at least partly (e.g. Goodall 1986, McLennan & Hockings 2016). Cannibalism does likewise occur, with unrelated infants a favourite target (e.g. Kirchhoff et al. 2018, Nishie & Nakamura 2018, Lowe et al. 2020). Moreover, in some habitats and to various degrees, chimpanzees exploit colonies of eusocial insects such as termites, ants and bees to consume imagos, brood or honey – often aided by tools (McGrew 1992, Pruetz et al. 2006, Sommer et al. 2017). Moreover, chimpanzees will sometimes ingest soil, in particular from termite mounds, as well as small stones, presumably to detoxify food or to aid digestion (Pebsworth et al. 2019). Self-directed coprophagy has also been observed, perhaps as a means to repopulate the gut flora.

Within this comparative framework, in the following, we sketch out the dietary portfolio of the Rubondo apes.

Measuring Food Supply

We supplemented our own data with those from other Rubondo researchers. The most detailed information comes from a 19-month study from Oct02 to Apr04 conducted by Liza Moscovice and colleagues (Moscovice 2006: 29–68, Moscovice et al. 2007). Here, 11 months of phenological monitoring from Feb–Dec03 relied on 9 randomly placed transects (300–500 m × 10 m) in a north-western, northern and central part of the island (Chapter 3). Chimpanzee diet, apart from occasional direct observations, was reconstructed via analyses of faecal samples, mostly collected from below fresh nests but also while following the animals (n = 147 samples, median 7/month, range 0–27). When chimpanzees eat fruit, they tend to swallow the seeds, rather than destroying or spitting them out (Lambert 1999). This allows identification of food plants in the faeces, typically to species level.

Extending this research, Klára Petrželková et al. (2010a), in 2008, employed infrared spectroscopy for nutritional analysis of faeces. The method is based on different reflections of infrared radiation by substances such as NDF (Neutral Detergent Fibre, i.e. hemicellulose, cellulose, lignin), ADF (Acid Detergent Fibre, i.e. cellulose, lignin), nitrogen free extract and fat.

Food Sources

As for the main food – fruit – it is useful to reiterate findings from the 2003 study presented in Chapter 3, which established that availability is similar across island sections. Trees grow more fruit in months with more rainfall (Pearson, n = 11, r = 0.734, p = 0.01), while this is not true for lianas (r = −0.059, p = 0.863) (Figure 5.8). Thus, variations in tree fruit availability are buffered by an abundance of liana fruit, which, even when not actively sought out, can be eaten opportunistically. This important macroscopic result is refined by infrared spectroscopy, which likewise reveals an absence of seasonal fluctuation in fruit quality in terms of NDF, ADF and nitrogen (Petrželková et al. 2010a) – a finding quite different from seasonal habitats such as Gashaka, Nigeria (Sommer et al. 2011).

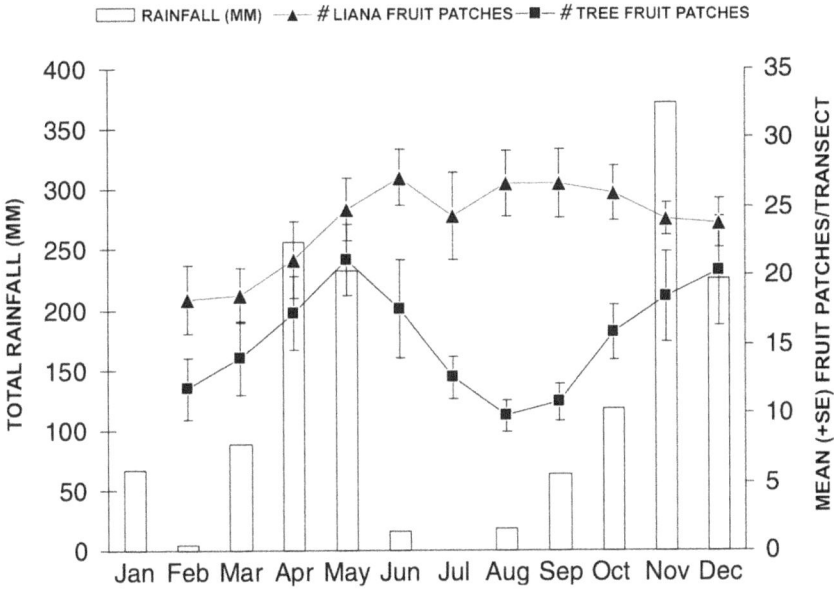

Figure 5.8 Mean (+/– SD) density of tree and liana fruit patches on Jan–Dec03 transects in relation to rainfall. (From Moscovice et al. 2007: 493, Figure 1)

To arrive at a picture of the Rubondo chimpanzees' plant food species, we combined records from 1966 to 2014 (excluding a preliminary list provided by Kiwango et al. 2005: 99–110, a revised version of which is found in Moscovice 2006). The compilation amounted to 56 different species (Table 5.3). Of the life-forms that provided food (n = 59), trees constituted 59.3%, lianas 25.4% and shrubs 10.2%. Of all identifiable plant food items (n = 50), fruits accounted for 94%, while the rest were leaf, gum and pith.

Based on the percentage of faeces that contained seeds, the most important food plants were as follows (Moscovice 2006: 68, Moscovice et al. 2007):

- 66% *Saba comorensis*, var. 1 (rubber vine);
- 58% *Ficus* (figs);
- 47% *Garcinia huillensis* (granite mangosteen);
- 26% *Saba comorensis*, var. 2 (rubber vine);
- 22% *Drypetes gerrardii*;
- 20% *Antiaris toxicaria*;
- 17% *Teclea nobilis*;
- 15% *Strychnos lucens*;
- 12% *Pseudospondias microcarpa*;
- 12% *Uvaria* sp.;
- 11% *Phoenix reclinata* (wild date palm).

Table 5.3 Chimpanzee food plants (n = 56 taxa) compared to transect specimens (%, 2012–14; cf. Table 3.2)

Family	Latin	Common name	KiSwahili (Zinza)	Transect specimens (%)	Life-form	Eaten	Evidence (O = observed, F = faeces)	Source
Mimosoideae	Albizia gummifera	Peacock flower	Mkenge, mchani mbao	0.9	Tree	Gum	O	3, 4
Bromelicaceae	Ananas comosus	Pineapple	Nanasi		Shrub	Fruit	O	1
Annonaceae	Annona senegalensis	Wild custard apple	Mchakwe, mtopetope	0.6	Tree	Fruit	F	3, 4
Moraceae	Antiaris toxicaria	Bark cloth tree, antiaris, sacking tree	Mkunde	7.4	Tree	Fruit	O/F	3, 4
Rubiaceae	Canthium lactescens	Hairy turkey berry, gummy canthium	(omusangate)	1.2	Tree	Fruit	F	3
Capparidaceae	Capparis erythrocarpos	Red-fruited, capter bush	(mtungu)		Liana	Fruit	F	3
Capparidaceae	Capparis tomentosa	Woolly caper bush	Mbada paka		Liana	Fruit	F	3
Polygalaceae	Carpolobia conradsiana	Carpolobia	(omugusa)	2.2	Tree	Fruit	O	3, 4
Ulmaceae	Chaetacme aristata	Basterwitpeer, doringolm	(omugunga)		Tree	Fruit	O	3
Vitaceae	Cissus quadrangularis	Veldt grape, devil's backbone, adamant creeper	(munia)	0.6	Liana	Fruit	O	3
Rutaceae	Citrus limon	Lemon	Limau		Tree	Fruit	F	3
Rutaceae	Citrus x sinensis	Orange	Machungwa		Tree	Fruit	O	1
Rubiaceae	Coffea eugenioides	Nandi coffee	Kahawa (omuhekaheke)		Tree	Fruit	F	3

Family	Species	Common name	Local name		Growth form	Part		
Combretaceae	*Combretum molle*	Velvet bush willow	Kagua, mdama, mlamam	0.1	Tree	Leaf	O	2, 3
Nyctaginaceae	*Commicarpus plumbagineus*				Perennial		O	1
Euphorbiaceae	*Croton sylvaticus*	Forest fever-berry	Msinduzi	5.8	Tree	Fruit	O/F	3
Cyperaceae	*Cyperus papyrus*	Nile grass, papyrus			Perennial	Pith	O	1
Sapindaceae	*Deinbollia fulvo-tomentella*		(omuhuna)		Liana	Fruit	F	3
Putranjivaceae	*Drypetes gerrardii*	Bastard white ironwood	Kihambie (omufi)	0.2	Tree	Fruit	O/F	3
Moraceae	*Ficus sansibarica*	Knobbly fig	Mkuyu	0.7	Tree	Leaf	O	3
Moraceae	*Ficus sp. (thonningii ?)*	Fig			Tree	Fruit	O	1, 3
Flacourtiaceae	*Flacourtia indica*	Batoka plum	Duruma, madungatundu	0.1	Tree	Fruit	F	3
Annonaceae	*Friesodielsia obovata*	Monkey fingers			Shrub, tree	Fruit	O	2
Guttiferae	*Garcinia buillensis*	Granite mangosteen	(esarazi, sharazi)	1.1	Tree	Fruit	O/F	3, 4
Tiliaceae	*Grewia bicolor*	White raisin, white-leaved raisin	(omukomako)		Tree	Fruit	F	3
Tiliaceae	*Grewia flavescens*	Sandpaper raisin, donkey-berry			Liana, shrub	Fruit	F	3
Malvaceae	*Grewia forbasii mast*			0.1	Tree	Fruit	F	3
Sapindaceae	*Haplocoelum foliolosum*	Northern galla-plum		1.4	Tree	Fruit	F	3
Malvaceae	*Hibiscus rostellatus*	Hibiscus			Shrub	Fruit	O	1
Acanthaceae	*Hypoestes verticillaris*	White ribbon bush			Herb	Fruit	O	1
Rubiaceae	*Keetia venosa*	Raisin-fruit keetia			Liana, shrub	Fruit	F	3

(*Continued*)

Table 5.3 Continued

Family	Latin	Common name	KiSwahili (Zinza)	Transect specimens (%)	Life-form	Eaten	Evidence (O=observed, F=faeces)	Source
Anacardiaceae	*Lannea fulva*		(murangalala)	0.1	Tree	Fruit	F	3
Lecaniodiscus	*Lecaniodiscus fraxinifolius*	River-litchi	Mbwewe, mkunguma	0.6	Tree	Fruit	F	3
Rhamnaceae	*Maesopsis eminii*	Musizi, umbrella tree	Msizi, muhunya, ndunga	0.1	Tree	Fruit	O	3
Anacardiaceae	*Mangifera indica*	Mango	Embe		Tree	Fruit	O	1
Sapotaceae	*Mimusops kummel*	Red milkwood	Mgambo	0.2	Tree	Fruit	F	3
Moraceae	*Morus mesozygia*	Black mulberry	Mkuzufunta (muchimbo mkimbi)	0.9	Tree	Fruit	F	3
Sapindaceae	*Pancovia turbinata*		(omutemelele)	17.2	Tree	Fruit	O/F	3
Dichapetalaceae	*Parinari curatellifolia*	Hissing tree, mobola plum, fever tree	(omunazi)		Tree	Fruit	O	3, 4
Palmae	*Phoenix reclinata*	Wild date palm	Mkindu		Tree	Fruit	F	3
Anacardiaceae	*Pseudospondias microcarpa*	African grape	Omubolu	0.8	Tree	Fruit	O/F	3
Icacinaceae	*Pyrenacantha sylvestris*				Liana	Fruit	O/F	3
Apocynaceae	*Saba comorensis, var. 1 and var. 2*	Rubber vine	Bungo, mabungo		Liana	Fruit	O/F	2, 3
Celastraceae	*Salacia erecta*				Liana	Fruit	O/F	3
Celastraceae	*Salacia leptoclada*	Sand lemon-rope	Kabamba, kasolia	0.2	Liana	Fruit	O/F	3, 4
Loganiaceae	*Strychnos lucens*				Tree, liana	Fruit	O/F	3

Family	Scientific name	Common name	Local name		Life form			
Sapotaceae	*Synsepalum brevipes*		Mchocha mke, mchocho jike	4.2	Tree	Fruit	O/F	3
Rutaceae	*Teclea nobilis*		Muzo (olulema)	3.1	Tree	Fruit	O/F	3
Menispermaceae	*Tinospora caffra*	Orange grape creeper			Liana	Fruit	F	3
Malvaceae	*Urena lobata*	Caesar weed, Congo jute			Shrub		O	1
Annonaceae	*Uvaria angolensis*		Mbola, mobola		Liana	Fruit	O/F	3
Annonaceae	*Uvaria welwetschii*				Liana	Fruit	O/F	3, 4
Verbenaceae	*Vitex doniana*	Black plum	Mfudu, mfuru, mfuu	2.6	Tree	Fruit	O/F	3
Malvaceae	*Wissadula periplocifolia*	White velvetleaf			Perennial		O	1
Sapindaceae	*Zanha golungensis*	Smooth-fruit zanha	(omwikaliwankobe)	0.4	Tree	Fruit	O/F	3

Source: 1 = 1966–68, 10 species (Kade 1967; and report to FZS, cit. in Müller & Anzenberger 1995: 72); 2 = 1994, 2 species (Müller & Anzenberger 1995: 37); 3 = 2002–04, 46 species (Moscovice 2006: 66f); 4 = 2012–14, 8 species (Msindai 2018). Scientific nomenclature follows Beentje (1994, 2002), Quattrocchi (2012), and internet sites (accessed 01–10 Apr 2015: aluka.org, plantzafrica.com, worldagroforestrycentre.org, zimbabweflora.co.zw, theplantlist.org, wikipedia.org/en, prota4u.org, plants.jstor.org, zambiaflora.com, treesa.org)

Thus, most faeces contained one or the other variation of *Saba comorensis*, a liana locally called *bungo* (plural *mabungo*) or rubber vine. While var. 2 has small fruits, mature fruits of var. 1 weigh 168 g (±35), and are amongst the largest fleshy fruits present on the island. The fruit has the appearance of an orange, because of its shape and hard orange-coloured peel (cf. Figure 3.8). When opened, its dozen or so fibre-covered pips reveal a texture similar to mango seeds. Incidentally, local people in Tanzania will often process the aromatic juice into a popular drink. Fruits of *Saba* are a major component of chimpanzee diet at other locales as well (e.g. Gombe, Tanzania, Goodall 1986; Mahale, Tanzania, Itoh 2002). To single out another important food species, *Ficus* seeds were likewise present in the majority of faecal samples – confirming findings elsewhere that they are, if not a preferred item, a major staple food (Harrison & Marshall 2011).

The chimpanzees also eat some plant types whose productivity was not monitored on transects. This includes tubers. On that account, during our 2012–14 study, we came across a partially eaten unidentified plant, its flesh resembling a watermelon in terms of colour and consistency, perhaps some sort of wild potato. Only recently was it recognised that chimpanzees, at least sometimes, consume underground storage organs of plants (Hernandez-Aguilar et al. 2007). Its role, even if marginal, in the diet of the Rubondo residents will need further exploration.

Several non-plant foods were also detected in diet remains, i.e. termites (*Microtermes, Odontotermes* sp.), ants (*Polyrhachis* sp.) and grasshoppers (*Homorocoryphus nitidulus vicinus*) (Moscovice 2006: 45, Moscovice et al. 2007). In addition, the chimpanzees were observed to consume vervet monkeys and sitatunga antelopes – cases of faunivory we revisit below.

Overlap between Food, Nesting and Transect Trees

To what degree do the Rubondo chimpanzees rely on the same tree types for food as well as nesting – and which species do they not utilize at all?

To explore overlap between food and nesting tree species, we compiled data for 2002–07 and 2012–14. This yielded a sample of 1,740 individually identified nesting trees belonging to 31 species. It turned out that 93.4% of nest specimens belonged to species also exploited for food, equivalent to 65.6% of all food-providing species.

Perhaps more interesting is the question if certain types of trees are avoided. For this, we did not focus on the absolute number of specimens, but on the number of species found on the 2012–14 transects. Thus, of 31 nest species, only 2 (6.5%) are not also found on the transect, while of 35 food species, 10 (28.6%) are not represented. From this we cannot conclude much more than that our transects, while random, are not extensive enough to catch all species important for chimpanzees. Still, we can discern that they are less selective of nest trees than food trees.

In a next step, we therefore explored the reverse question, i.e. whether certain trees are *not* used for food, despite the fact that they serve as as sleeping places. Indeed, the three most prominent of these tree species amounted to a quarter of all transect specimens:

- *Dichapetalum stuhlmannii* 13.0%;
- *Alchornea hirtella* 7.6%;
- *Maerua duchesnei* 3.6%.

We do not know what causes this discrepancy between being used one way (nesting) but not the other (food). Potentially, some nesting trees are avoided as food sources, because they have toxic properties. Thus, the *CRC World Dictionary of Medicinal and Poisonous Plants* (Quattrocchi 2012) lists *D. stuhlmannii* as "very poisonous to stock, to man, cattle and goats" – which explains the common name "goat killer". However, vice versa, several if not all preferred food species such as *G. huillensis* and *D. gerrardii* likewise possess toxic qualities. In particular (nomen est omen), the sap of *Antiaris toxicaria* can cause cardiac arrest – albeit the fruits are edible, while the leaves of *Strychnos lucens* contain strychnine. Similar to what we discussed above with respect to potential medicinal properties of nesting tree species, it will be difficult to extract causal factors from a profusion of plant properties that range from poisonous to remedial – not least because of the famous insight of Paracelsus: "Nothing is without poison; only the dose makes a thing not a poison."

Cross-site Comparison

There is considerable similarity between the portfolio of plants consumed by the Rubondo apes and chimpanzees elsewhere. The percentages of overlap for selected sites in terms of trees (T) and lianas (L) are as follows (Moscovice 2006: 110, Table 5.4):

- Gombe, Tanzania, species T 43%, L 40%, genera T 74%, L 63%;
- Mahale, Tanzania, species T 43%, L 30%, genera T 82%, L 69%;
- Budongo, Uganda, species T 53%, L 30%, genera T 82%, L 75%;
- Lope, Gabon, species T 15%, L 0%, genera T 76%, L 69%;
- Bossou, Guinea, species T 11%, L 5%, genera T 56%, L 63%.

This translates into an overall average species overlap of 27.0% and genera overlap of 70.9%. Unsurprisingly, the similarity is greater for the geographically closer East African communities compared to those from the Atlantic side of the continent, given region-specific differences in plant cover.

The important distinguishing characteristic between the Rubondo habitat and other sites is therefore not so much plant food diversity, but the non-seasonal availability of plant-derived food stuffs.

To conclude, the Rubondo inhabitants can rely on a stable high-quality diet, without significant seasonal variation or variation between individual years, as short-term intra-annual fluctuations are buffered by the availability of nutritionally rich liana fruits.

Gut Feelings: Parasitism and Mutualism

Never mind what chimpanzees can 'handle', whether weaving branches to construct nests or plucking morsels to curb hunger. Ultimately, survival depends on those 90% of cells in the body, that are not 'chimpanzee' (or human or monkey, for that matter), but uni- or multi-cellular symbionts of either the harmful kind (parasites) or the kind kind (mutualists). In particular, a wide array of microbiota – e.g. bacteria, viruses, archaea, eukaryotic microbes – interact with their host in complex connexions, thus constituting a 'metaorganism' (Bosch & McFall-Ngai 2011).

Apart from reigning into matters of individual life and death, symbionts, in an academic way, can enlighten us about how evolution came about and how ecosystems function. Traditionally, phylogenetic relationships were explored via comparative anatomy. Studying similarities and differences allows life-forms to be categorised into clades, i.e. groups that comprise all members which share a common ancestor. In addition, evolutionary pathways can be reconstructed by investigating symbiotic interactions between members of different species (review in Paracer & Ahmadjian 2000), such as embodied in parasitism or mutualism.

In parasitism, the host is harmed while the freeloader gains some benefit by living on it or in it (think ectoparasites such as fleas or lice that feed on those that harbour them, and endoparasites such as tapeworms or nematodes which consume some of the host's food). In mutualism, both organisms benefit from each other (think microorganisms that populate their hosts' digestive tracts, constituting the gut flora). Thus, in addition to comparative anatomy, information about the coevolution of hosts and their symbionts allows an independent evaluation of inferred phylogenies.

In a more applied way, understanding parasitic and mutualistic relationships also enables us to infer the influence that local ecologies have upon these associations (Huffman & Chapman 2009). In that respect, translocations of organisms into a novel environment are of particular interest, as endemic symbionts may begin to colonize the released animals, while native animals may be exposed to hitherto unencountered symbionts. The transmissions of novel parasites can have grave effects on naive hosts (Viggers et al. 1993, Cunningham 1996) – as evidenced by outbreaks of communicable diseases that devastated wild populations, including those of great apes (Köndgen et al. 2008, Kaur et al. 2008).

Both lines of inquiry – about phylogenetic pathways as well as the role of symbionts in translocations – have been pursued for the Rubondo apes. Such research can help us to assess potential risks of introducing large mammals into a foreign ecosystem and also if and how the life-history of ex-captives still echoes in their wild-living descendants.

Gastrointestinal Parasites

The Rubondo chimpanzees were the subject of several studies of gastrointestinal symbionts led by Klára Petrželková and Michael Huffman (pinworm morphology: Hasegawa et al. 2005; general parasite fauna: Petrželková et al. 2006, 2010b, Petrášová et al. 2010; *Blastocystis*: Petrášová et al. 2011; entodiniomorphid ciliate *Troglodytella abrassarti*: Pomajbíková 2010a, Vallo et al. 2012; *Balantoides coli*: Pomajbíková et al. 2010b; *Trichuris*: Dolezalova et al. 2015).

Trying to infer the source of parasitic infections is complicated, because the founders might have been given anti-protozoal and anthelminthic drugs during their captivity and shortly before release (Markus Borner, cit. in Petrželková et al. 2010b) – albeit it is not certain that such treatment took place. Certainly, the chimpanzees destined to be released were fed Resochin syrup, a prophylaxis for malaria (Grzimek 1971).

Our summary of findings on parasites that colonize the stomach and intestines (Petrželková et al. 2010b) is based on two sets of faecal samples (first, Mar97–Oct98, collected by FZS staff Paula and Johan Robinson and assistants; second, Oct02–Dec05, collected by Liza Moscovice, Mwanahamisi Issa Mapua, Klára Petrželková, Michael Huffman; cf. Table 2.3). As the samples could not be attributed to individuals, identical animals were theoretically sampled more than once on a particular day. Results are therefore based on so-called 'day samples' about parasite prevalence (percent of samples collected on a given day harbouring a specific species/genus) and parasite richness (number of unique parasite species/genera recovered from the day sample).

None to six parasite taxa were found in day samples. In the following list, the values behind each taxon indicate the percentage of positive day samples during 1997–98 (n = 89 faecal samples collected on n = 58 individual days) resp. 2002–05 (n = 179 samples collected on n = 78 days).

Protists:

- *Troglodytella abrassarti* 36.2 / 53.8;
- *Entamoeba* spp. 48.3 / 47.4;
- *Lodamoeba buetschlii* 10.3 / 1.3;
- *Chilomastix mesnili* 3.4 / 0.0.

Nematodes (roundworms):

- Strongylida fam. gen. (threadworm) 9.0 / 3.8;
- *Strongyloides* spp. (threadworm) 41.3 / 16.7;
- *Enterobius anthropopitheci* (pinworm) 8.6 / 17.9;

- *Trichuris* sp. (whipworm) 8.6 / 5.2;
- Ascaridida fam. gen. 5.2 / 0.0;
- *Protospirura muricola* 3.4 / 10.3;
- *Anatrichosoma* sp. 1.7 / 1.3;
- *Subulura* sp. 0.0 / 1.3.

Thus, both sample periods led to more or less identical findings. With the exception of threadworm (*Strongyloides*, Strongylida), there was no difference in parasite prevalence between the two periods, and neither was prevalence or parasite richness different between seasons (dry, transient, wet). Generally, the parasite fauna of the Rubondo chimpanzees is similar to those of native populations (review in Huffman & Chapman 2009). The most notable difference is a comparably very low prevalence of *Strongylida* on Rubondo.

Conceivably, if the releasees received anthelmintics, this could have eliminated or significantly decreased earlier strongylid nematode infections. Moreover, the comparably very large ranging opportunities (Chapter 4) may reduce the likelihood of reinfection. Still, Rubondo samples showed a considerable prevalence of pinworm and whipworm. Eggs of these parasites are thick-shelled, which hampers their eradication in captive facilities. Ergo, despite the potential application of anthelminthics, the founders may have maintained those parasites since infancy (chimpanzee specific *Enterobius anthropopitheci*) or in the case of *Trichuris* sp., acquired them in captivity.

Some parasites such as the small *Ascaris*, found at very low prevalence – were likely incidentally transmitted from other mammals on the island, given that their eggs are deposited in faeces and soil, and that plants with eggs on them infect those who ingest them. These probably acquired parasites also include three genera of nematodes, *Protospirura*, *Anatrichosoma* and *Subulura*, all of which were not previously described for chimpanzees (Petrželková et al. 2006). The presence of *Protospirura* adults in Rubondo apes very likely indicates that they were acquired via the consumption of grasshoppers, which serve as intermediate hosts. While the consumption of rodents and vervet monkeys (see below) may lead to the acquisition of the other two types – *Anatrichosoma* and *Subulura* –, these were detected in very low frequency only. This raises the possibility that they are examples of pseudo parasitism, in that eating the hosts could have led to deposition of their eggs in chimpanzee faeces.

None of the parasites are likely to constitute a serious health risk to chimpanzees. In particular, most representatives of the amoebas of genus *Entamoeba* (the exception being *E. histolytica*) are considered harmless commensals rather than parasites. Still, some intestinal protozoans and helminths can be pathogenic, inducing diarrhoea, gastric pain, anaemia, and pulmonary problems, with hyperinfections causing death in immunocompromised hosts (Mehlhorn 1988). Having said that, we have no indication that chimpanzee health is adversely affected, whether by parasites the founders maintained from since they were born, or from having acquired them in captivity, or from transmission after being released onto Rubondo.

At several sites, chimpanzees are known to fold and swallow bristly leaves which, in their unchewed condition, pass undigested through the intestines. This behaviour is considered to reflect a mechanism of self-medication in that it interrupts the reproductive cycles of parasites such as *Oesophagostomum stephanostomum* (Huffman et al. 1996, Huffman & Caton 2001, Fowler et al. 2007). No such evidence has so far been described for the Rubondo chimpanzees, albeit routine macroscopic inspections would be necessary to detect leaf-swallowing.

Troglodytella abrassarti

A follow-up investigation focused on a gastrointestinal protist that was identified during both sampling periods, *T. abrassarti*. This entodiniomorphid ciliate, rather than being parasitic, is a mutualistic symbiont supporting fermentation in the large intestine of herbivorous mammals such as elephants, rhinoceroses, horses, hippopotami, and various rodents (for literature, see Profousová et al. 2011, Vallo et al. 2012). These protists enable the hosts to utilize dietary fibre otherwise not digestible by animal hydrolytic enzymes.

Among primates, only African great apes are colonized by various types of entodiniomorphid ciliates. Early humans, aided by these protists in the same way as gorillas and chimpanzees still are, were perhaps also able to digest large quantities of plant material. However, given that modern humans lack these symbionts, their elimination might be connected to changes in diet once fire began to play a major role in food preparation (Wrangham 2009).

Because members of the family *Troglodytellidae* are specific to non-human African great apes, this limits the dispersal ability of the mutualist, offering the opportunity to study the codivergence in the host and its evolutionary partner, the ciliate. Similarly coupled phylogenetic history has been demonstrated in lice, where splits of lineages leading to extant ape species involved the parallel diversification of sister species of the lice genera *Pediculus* and *Pthirus* (Reed et al. 2007).

The phylogenetic study which includes samples from Rubondo (details in Vallo et al. 2012) relied on *T. abrassarti* as the model organism. The protist is typically harboured by chimpanzees, but also occurs in gorillas and probably bonobos. The investigation utilized faecal samples of the Rubondo apes (n = 2) plus those from all four subspecies, and also bonobos, i.e. *Pan troglodytes schweinfurthii*: Kalinzu, Uganda (n = 3 samples); Kyambura, Uganda (n = 1); Budongo, Uganda (n = 1); Ugalla, Tanzania (n = 2); *P. t. troglodytes*: Loango, Gabon (n = 1); Goualougo, Congo (n = 2); *P. t. ellioti*: Gashaka, Nigeria (n = 2); *P. t. verus*: Taï, Ivory Coast (n = 1); Bossou, Guinea (n = 1); *P. paniscus*: Wamba, DR Congo (n = 2).

One to three cells of *T. abrassarti* were picked from the samples that had been conserved in ethanol or RNAlater. Genomic DNA extracted via the

variability of amplified sequences of small subunit ribosomal DNA (SSU rDNA) was assessed and related to the geographical localities. Phylogenetic relationships were inferred via a selected spectrum of entodiniomorphid and vestibuliferid ciliates occurring in the gastrointestinal system of ruminants and perissodactyls.

In contrast to previous research on *Troglodytella*, the comprehensive spatial origin of samples allowed the detection of genetic variability potentially inherent within a wide geographic distribution of the host species, given that the natural range of chimpanzees stretches over 5,000 km from Senegal to Tanzania.

The study established a low variability of *T. abrassarti* sequences throughout the *Pan* range. This suggests the conspecificity of all sequenced ciliates, implying a historical overlap of the distribution ranges. However, the categorisation of distinguishable types still follows a pattern of chimpanzee evolutionary units (Figure 5.9). Thus, the analyses established a difference between a Type I and a Type II of *T. abrassarti*, with the former probably constituting a derived form. Type I was found to be restricted to the West African subspecies *P. t. verus*, whereas Type II is exclusive to the three other subspecies in Central and East Africa. The identity of *T. abrassarti* from Nigerian *P. t. ellioti* and Central African *P. t. troglodytes* suggests their close

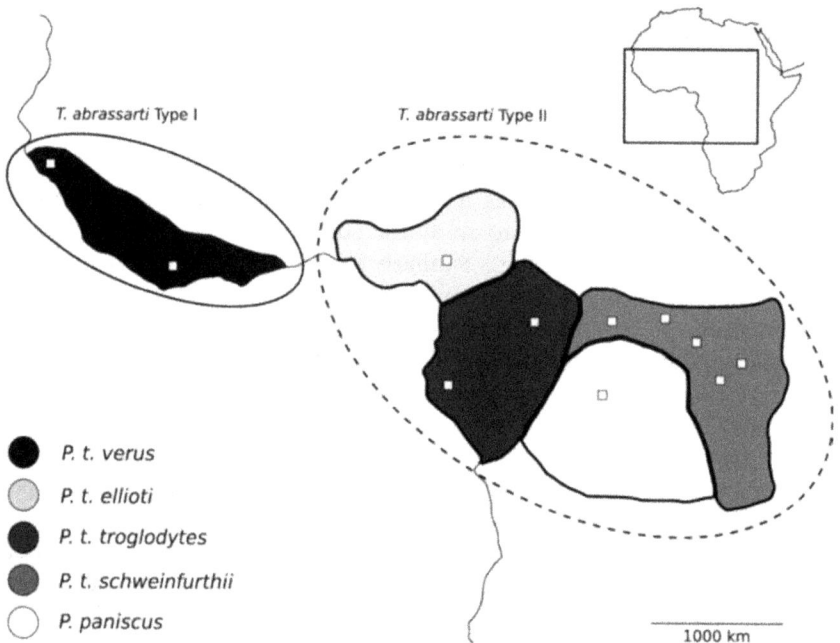

Figure 5.9 Grouping of obtained DNA sequences of the ciliad *Troglodytella abrassarti* in chimpanzee subspecies and bonobos. Sampled localities denoted as white squares. (From Vallo et al. 2012: Figure 1)

evolutionary relationship, although this conflicts with data, which group *P. t. ellioti* with *P. t. verus* (Bjork et al. 2011, Gonder et al. 2011). The fact that Type II is also found in bonobos suggests that this is an ancestral type dating back to the common *Pan* ancestor.

This relatively simple picture is tied to what seems strict host specificity – a finding quite different from the mentioned scenario of parasitic sucking lice of primates, which have coevolved with their hosts over at least 25 million years (Reed et al. 2007). The head and body lice of chimpanzees and humans last shared a common ancestor roughly 6 mya, and their divergence is con-temporaneous with that of their hosts. However, humans host two different genera of lice, one shared with chimpanzees and another one shared with gorillas. Thus, a common *Pthirus* ancestor probably switched from gorillas to humans, and evolved into two extant species *P. gorillae* in gorillas and *P. pubis* in humans. Evolutionary events in relation to the lice of African apes therefore represent a complex mixture of cospeciation, parasite duplication, parasite extinction, and host switching. The history of *T. abrassarti* is much more parsimonious, in comparison.

So, what about the Rubondo apes and *T. abrassarti*? While the island is closest to the range of East African chimpanzees, the *T. abrassarti* found in the descendants of the introduced apes is not akin to the East African sub-species *P. t. schweinfurthii*, but akin to the Type I found in the West African *P. t. verus*. These findings corroborate the results of our own genetic analy-ses (Chapter 2). However, the congruence is only partly, given that the stock of the Rubondo apes, apart from *P. t. verus*, also contains genes from the Central African subspecies *P. t. troglodytes*. Potentially then, if both types colonize ape guts, Type I is able to outcompete Type II – an explanation, which, at this stage, is not more than speculation.

Still, what is notable – apart from the type distribution according to sub-species variety – is the fact that these highly host-specific mutualistic sym-bionts did survive in the guts of wild-born animals that were kept for years in captivity. Interestingly, when a related study (Pomajbíková et al. 2010) screened captive chimpanzees for *Troglodytella*, ciliates of this genus were detected in only 13 of 23 colonies, perhaps because of the administration of chemotherapeutics or antiparasitic or antibiotic drugs. However, even if some of the Rubondo founders were subjected to similar treatments, it would have been sufficient for a single individual still harbouring *Troglodytella* to subsequently reinfect its comrades.

Gut Microbiome

A similar result – namely the detection of a trait that can be traced back to captivity – transpires from another investigation of gut symbionts, which in this case focussed on their whole diversity, i.e. the gut microbiome.

The assemblage of microbiota forming the gut microbiome benefits from the shelter and nourishment provided by their hosts. Conversely, hosts gain protection against germs, and microbiota also produce vitamins or break down food to release energy – such as when cellulose-digesting protozoans or bacteria help herbivores to digest plant matter (reviewed in Clayton et al. 2018). Newborns first acquire microbiota during the birthing process. The gut microbiome is therefore partially vertically inherited (Mueller et al. 2015), albeit its diversity changes throughout life, not least through horizontal dispersal from other group members.

As a consequence, gut microbiomes, including those of primates, vary according to host species, biogeography, climate, diet, sociality, and disease state (cf. Clayton et al. 2018). Using 560 faecal samples collected from wild chimpanzees across their range, Bueno de Mesquita et al. (2021) assessed the influence of such environmental factors. The study found a high degree of regional specificity in the microbiome composition, which was associated with host genetics, available plant foods, and potentially with cultural differences in tool use, which affect diet. Thus, while regional differences are undetectable in industrialized human populations, chimpanzee gut microbiomes are far more variable across space. This suggests that developments in technology decoupled humans from their local environments, diminishing regional variations that could have been influential during human evolution.

Similarly and unsurprisingly, captive animals will also often experience a loss of the native gut microbiome. A variety of factors are involved in this process (Campbell et al. 2020; see also Song et al. 2013, Moeller et al. 2013, 2016, Hicks et al. 2018). Thus, medication such as antibiotics can cause declines in certain microbiota, creating chances for novel taxa to colonize a host. Dietary shifts also play a role. Moreover, due to interactions and close contact with caretakers, there is ample opportunity for cross-species transmission from humans. Finally, as for captive primates in particular, they are typically provided with cultivated fruits and vegetables high in soluble sugars, higher in polysaccharides but lower in fibre compared to the wild. This leads to a corresponding proliferation of taxa specialised in the degradation of carbohydrates and a lower abundance of taxa specialised to digest fibrous material.

While a number of studies have compared captive versus wild primates (e.g. Tsukayama et al. 2018), there has been no research on the gut microbiomes of free-living individuals whose recent ancestors were kept captive. Given that the Rubondo chimpanzees remain geographically isolated from other apes, they provide an ideal opportunity to investigate if and how captivity and inheritance affect the gut microbiome – including the interesting question whether captive effects can be 'reversed' following release into the wild and ensuing consumption of wild-growing foods.

Through research led by Catryn Williams (Williams 2020, Williams et al. in prep.), we compared the Rubondo chimpanzee gut microbiome with three other populations, i.e. two native and wild living communities as well as a 'pseudo-population' of zoo animals:

- *Captive population*. We investigated faecal samples of 9 chimpanzees kept at UK zoos in Colchester, Twycross and Whipsnade who were all captive-born and had spent their entire lives in captivity.
- *Native population at Gashaka, Nigeria*. We analysed 12 faecal samples from the *P. t. ellioti* community of Kwano living in north-eastern Nigeria in Gashaka Gumti National Park (altitude 583 m), which numbers about 35 individuals, occupying a home-range of about 30 km². Between 2000–12, mean minimum temperature was 20.8°C, mean maximum 31.9°C, and annual mean rainfall 2,021 mm (range 1,683–2,337 mm). As for seasonality, 5 months of very little or no rainfall are followed by heavy downpours from Apr–Nov constituting 97% of all precipitation. The chimpanzees inhabit a mountainous mosaic habitat consisting of 22% open woodland-savannah and 78% closed forests. Their diet is mostly fruit, while they also prey upon colonies of social insects, to harvest honey and ants (Sommer & Ross 2011).
- *Native population at Issa, Tazania*. We investigated 20 faecal samples from the *P. t. schweinfurthii* community of Issa living in Tongwe East Forest Reserve, western Tanzania (altitude 1,100–1,800 m), which numbered about 29 individuals, ranging over about 55 km². The chimpanzees inhabit an open, mosaic habitat, which – like Gashaka – is very seasonal. A wet season (Nov–Apr, when all rain falls, ~1,250mm) is followed by a distinct dry spell (May–Oct). The landscape is a miombo woodland, characterised by strips of riparian forests separated by broad valleys, and includes patches of swamp, grassland, and rocky outcrops. The diet is mostly fruit, supplemented by termites and ants (Piel et al. 2017).
- *Introduced population on Rubondo, Tanzania*. We analysed 20 faecal samples from the hybrid *P. t. verus* x *P. t. troglodytes* community (Chapter 2). To briefly recap context, the population of Rubondo Island National Park (1,134–1,486 m altitude; 237 km²) was estimated at 35 individuals in 2014, ranging over most of the island. The habitat consists of evergreen and semi-deciduous forest. There are two wet seasons (Mar–May, Oct–Dec) and one dry season (Jun–Aug). The Rubondo residents are free-living and self-sustaining, with a diet consisting of more than 90% fruit. The 16 founders, prior to their release in 1966–69, had spent 4–9 years in captivity in European zoos and circuses – in solitary confinement or being housed with other apes or humans. All founders were sourced at a young age from the wild in West- and Central Africa. Faecal samples were collected between 2012–14, when the population consisted of 1st and 2nd generation individuals.

Because previous research has highlighted that diet will cause the gut microbial community to shift in composition, we hypothesized that the current Rubondo chimpanzees would have a microbiome that is more similar to native communities than their captive counterparts – meaning, that we expected them to have reverted to a more 'natural' array of gut microbes.

The principal method relied on the extraction of genomic DNA from faecal samples (16S rRNA gene sequencing; details in Williams et al. in prep.). Given that it is difficult to define bacterial species (Baltrus 2016), the classical solution is to sort the reads into Operational Taxonomic Units (OTUs). In OTU picking, a sequence similarity cut-off is decided such that any reads that share at least 97% of their DNA sequence are grouped into one OTU and treated as one bacterial 'species'. Representative sequences are then picked for each OTU, and OTUs that share 95% sequence similarity are treated as a 'genus', and so on.

The diversity within a single sample is called alpha diversity, broken down further into richness (numbers of taxa) and evenness (the distribution of cell counts across these different taxa). Beta diversity refers to diversity between samples, or, how similar or dissimilar they are to each other. The first of two microbiome-specific measures is 'unweighted Unifrac', which uses a phylogenetic tree generated from the unique DNA sequences present across all samples to account for their taxonomic relationships. The second is 'weighted Unifrac', which likewise requires a phylogenetic tree and additionally takes into account the relative abundances of each unique sequence within a sample. Principal Coordinate Analysis (PCoA) plots visualise diversity.

In our specific research, differentially abundant taxa were filtered out to leave a 'core' microbiome common to both native communities to account for population-level variation. Likewise, OTUs present in fewer than five samples from captives were removed to account for individual differences. Samples for each population category were collapsed into a single representative sample.

When comparing alpha diversity, we found that the captive chimpanzees harboured greater numbers of bacterial taxa, a significant difference compared to the Issa and Rubondo apes (non-parametric two-sample t-test using 999 Monte Carlo permutations, $p = 0.001$ and $p = 0.048$ respectively), but not quite significant compared to Gashaka ($p = 0.054$). Similarly, captive baboons have higher alpha-diversity than wild-living counterparts (Tsukayama et al. 2018). While this contrasts with reports describing loss of microbial species diversity in captive primates (Clayton et al. 2018) and in Western human populations compared to non-Western groups (Yatsunenko et al. 2012). The conflicting results, more than anything, suggest that high alpha diversity is not indicative of increased host fitness in a given environment (Tsukayama et al. 2018).

When comparing beta diversity, we were surprised to find that the Rubondo chimpanzees clustered with the captives and away from the two native communities – a finding demonstrated by weighted Unifrac distances, (Figure 5.10a), with an even greater separation seen in unweighted Unifrac (Figure 5.10b).

To further test whether samples from within the same population were more similar to each other than to others, weighted Unifrac distances were

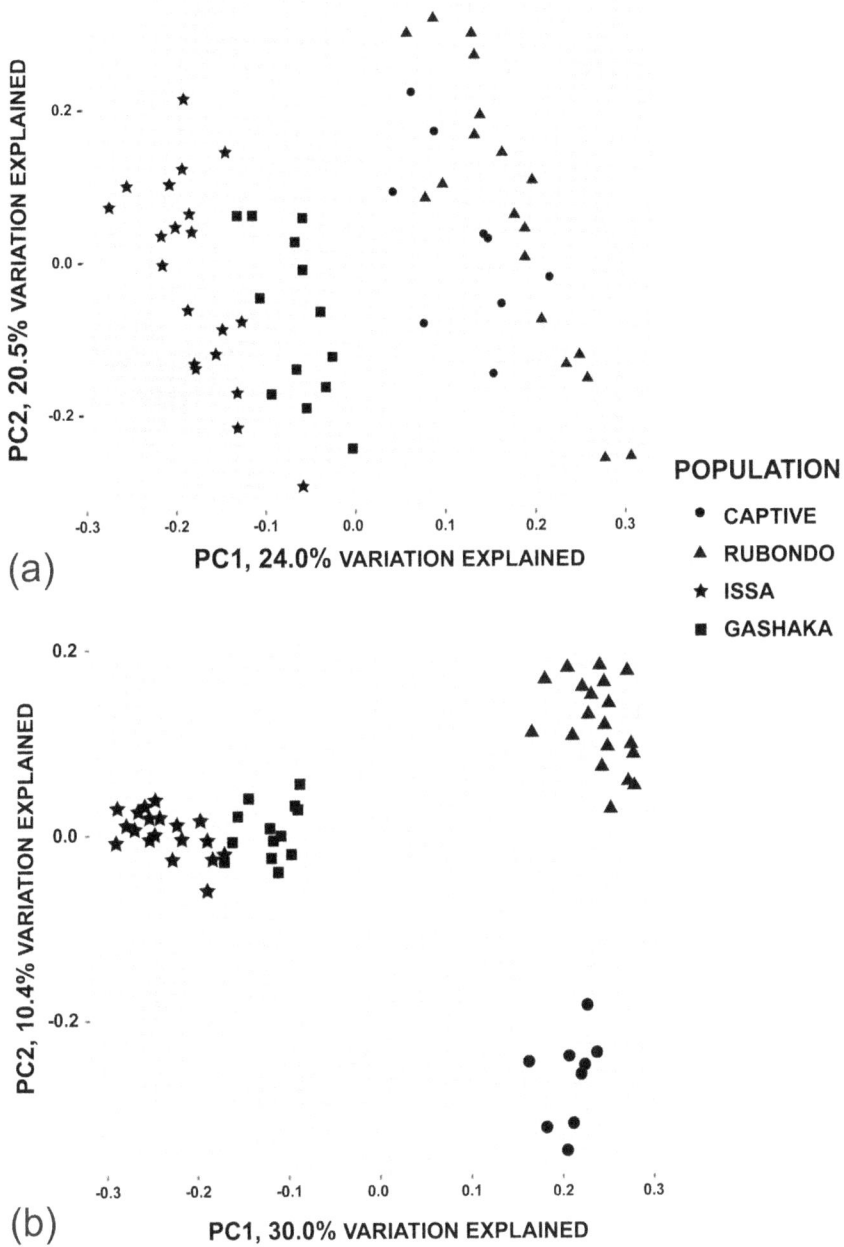

Figure 5.10 Quantification of beta diversity in chimpanzee microbiomes via PCoA plots. (a) Weighted Unifrac distances. (b) Unweighted Unifrac distances. (Modified after Williams 2020)

sorted into ten groups, based on which populations the two samples used to create each distance value came from. The samples were further sorted into 'within-population' and 'between-population' groupings (Figure 5.11). Unsurprisingly, distances were significantly lower between samples from the same site (two-sample t-test with Bonferroni correction, $t = -21.2$, $p = 0.001$), indicating that population membership significantly determines gut microbiome composition. Of the between-population groupings, the mean distance between the Issa and Gashaka was significantly smaller than the mean distances of all other between-population groupings (two-sample t-test, $p < 0.05$ for all comparisons), suggesting that the two native populations host the most similar gut microbiomes. The next most-similar grouping was that between Rubondo and captive chimpanzees, with a distance significantly lower than those between Rubondo and Gashaka or between Rubondo and Issa ($p = 0.02$ and 0.01 respectively). This, again, suggests that the Rubondo apes have gut microbiomes more similar in composition to those of the captives than to natives. The greatest distances existed

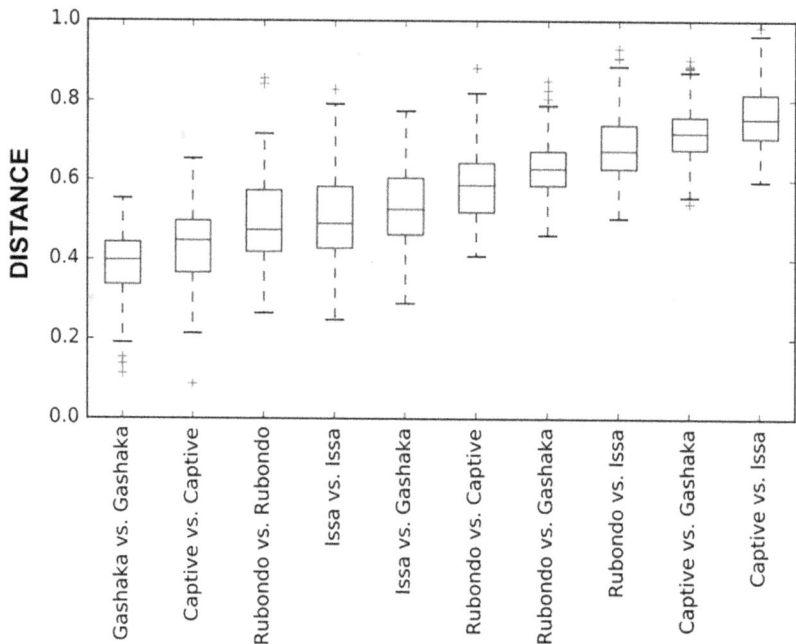

Figure 5.11 Boxplot of weighted Unifrac distances for chimpanzees, based on pairwise-calculated distance values. For example, the distance value calculated between two Gashaka chimpanzees is placed in the 'Gashaka vs. Gashaka' grouping, whereas that between a Gashaka and Issa chimpanzee is placed in the 'Issa vs. Gashaka' grouping. Boxes are sorted by mean, indicated by the line within each box. (Modified after Williams 2020)

between captive chimpanzees and those at Issa and Gashaka (p = 0.01 in both cases).

The subsequent founders of the Rubondo population were taken into captivity as infants, when gut microbiomes are highly susceptible to colonization to new bacteria (Walter & O'Mahony 2019). Because of this, they likely ended up having microbiota similar to those of our captive chimpanzee samples. The transition was probably catalysed by administered medications (Campbell et al. 2020) and over time, a new microbiome developed and stabilised, influenced by both diet and proximity to human caretakers, visitors and other captive animals.

However, what might explain the surprising finding that a microbiome acquired in captivity persists through multiple generations after release into the wild?

The gut microbiome is typically specialised for life in the lower intestines (Moeller et al. 2016), providing protection against colonization of the host by others (Round & Mazmanian 2009). More importantly, the persistence of captive-like features in the guts of the Rubondo residents was probably enabled due to a lack of native conspecifics in the habitat as a source to acquire novel taxa. This assumption is corroborated by the fact that sympatric chimpanzees and gorillas harbour similar gut microbiomes (Moeller et al. 2013). Therefore, we suspect that ex-captive apes have the potential to acquire a more typical wild gut microbiome if released into a habitat containing other great ape species. This assumption could be tested if faecal samples of animals destined for release (Chapter 6) are conserved and compared against samples from sometime after they have been exposed to the wild.

At the same time, the congruence between the native communities at Gashaka and Issa may be partly explained by their similar 'wild' diet, different from that in captivity. At Issa, at least 77 plant species are consumed, with the genera *Ficus*, *Garcinia* and *Saba* being the top preferred taxa (Piel et al. 2017), while at Gashaka 53 plant species are eaten, including those from the genera *Antiaris*, *Leea* and *Malicantha* (Hohmann et al. 2006). Diets of wild-living chimpanzees (see above) are dominated by fruits that are richer in fibre and lower in carbohydrates than captive foods, at times complemented by animal prey – sources not readily available to captives.

But, while the Rubondo menu overlaps with 15 plants from Issa and 16 plants from Gashaka, specifically *Saba*, *Garcinia*, *Ficus*, and *Antiaris*, a 'wild diet' is obviously not enough to revert a microbiome assembled in captivity and to instead bring about a more 'typical' free-living chimpanzee gut microbiome.

Nevertheless, the Rubondo apes seem to be in good health, opening up the possibility that the specific composition of a gut microbiome does not reflect strict adaptations to particular environments, or at least that the composition embodies considerable versatility. Clearly, the puzzling findings for the Rubondo apes indicate that we still have much to learn about gut microbiomes.

Klára Petržalková – on Rubondo 2003–08

Figure B5.1 Klára Petržalková during fieldwork. (Photo: David Modrý)

After completing my PhD on the ecology of bats, I pursued a career in primatology. I came to work on the island after Michael Huffman offered me an opportunity for post-doctoral studies. Rubondo thus inspired my passion for primate parasitology, for Africa and great apes. My accommodation was rather basic, we used also camps in the forest, with no running water and very limited electricity, which meant that we had to re-charge our radios at park headquarters. It was also a considerable challenge to find the chimpanzees. Numerous times we were chased by elephants, even while riding motorbikes. That was quite frightening at the time, but I now remember it fondly, and likewise how I made lifelong friends with some rangers. Our Tanzanian research assistant Mwanahamisi Issa secured a Czech governmental scholarship and could complete her master and PhD studies with our team in Brno, Czech Republic. I would also like to pay tribute to our colleague, keen Slovak naturalist Lucia Bobáková, who worked with us on the island and shortly after passed away very young due to cancer. I really enjoyed working on Rubondo, observing the changing of the seasons, hiking through its forest with its bountiful fruits, while tracking ghostly ape shadows.

Panthropology: Cultural Traits of the Rubondo Apes

Academics who explore human societies and their local customs of how to obtain, prepare and ingest food, or regional expressions of rituals and social conventions are called anthropologists. However, 'anthropology' recently spawned a new subdiscipline, aptly named 'cultural panthropology' (Whiten et al. 2003): the exploration of varied lifestyles in our closest living relatives, i.e. apes of the genus *Pan*.

Indeed, bonobos, but in particular chimpanzees, display an astounding degree of behavioural diversity likely unrivalled amongst non-human animals (Hohmann & Fruth 2003, McGrew 2004). As we have mentioned (Chapter 4), each study community displays its own mosaic of traits related to customs, communication, aggression, meat-acquisition, tool use or self-medication (Sommer & Parish 2010). This degree of plasticity is perhaps not surprising, given that *Pan* and the likewise very flexible *Homo* shared a common ancestor until about 4–6 million years ago.

A ground-breaking review (Whiten et al. 1999) compiled the behavioural patterns at six long-term chimpanzee research sites (Bossou, Guinea; Kibale and Budongo, Uganda; Taï, Ivory Coast; Mahale and Gombe, Tanzania). Behaviours for which ecological explanations seemed plausible were carefully discerned from a couple of dozen traits customary or habitual among some groups that did not reflect environmental constraints. These arbitrary traits were classified as purely 'cultural variants'.

The perhaps best-known example is the cracking of nuts with hammers of stone or wood. Apart from an odd report for Cameroon, the technique is restricted to Taï in West Africa, despite an abundance of nuts and potential tools elsewhere. The practice is probably neither genetically determined nor a corollary of particular environmental conditions. Instead, the N'Zo-Sassandra River in Ivory Coast seems to constitute a 'cultural boundary' between nut-cracking chimpanzees on the one bank and those who do not display this habit on the other bank, despite the obtainability of nuts and tool materials.

Of particular interest for the debate about ape culture are 'arbitrary' behaviours not tied to resource extraction, but which instead constitute inventions that somehow spread through a particular community (Whiten & McGrew 2001). For example, at most sites, but not in Taï and Bossou, chimpanzees engage in 'leaf-groom', i.e. they pick up one or more leaves, peer at them closely, manipulate them most intently, sometimes while lip-smacking. Similarly, 'leaf-clip' is performed at all sites but Taï and Bossou, in that the animals produce a conspicuous sound by ripping off a leaf, an act that seems to gain them attention. There is also the 'rain-dance', performed everywhere except at Bossou, when, at the start of cloudbursts, several adult males perform vigorous charging displays, coordinated or in parallel, including ground slapping, beating of buttresses, dragging of branches and pant-hoots.

Other well-described arbitrary traits are the positions individuals hold during mutual grooming. At Taï, Mahale and Kibale, apes will face each other while sitting on the ground, and each will fully extend the free arm overhead and clasp the partner's hand. It is hard to conceive an environmental pressure that would induce chimpanzees to attain this peculiar A-frame style of grooming. Yet, while Gombe is geographically close to Mahale and Budongo close to Kibale, the overhead hand-clasp is not observed there.

Similarly, chimpanzees at Gashaka, Nigeria, developed a sophisticated plant-based tool kit with which they prey upon colonies of honeybees, stingless bees, arboreal ants and army ants. However, they never employ the implements to fish for termites, albeit these are readily available and 'fishable', like at other sites (Sommer et al. 2017). This conspicuous lack of a particular kind of insectivory might be akin to a 'food taboo' observed in human communities, thus constituting a social marker that generates a sense of group identity (Sommer & Parish 2010).

Compared to other sites, our data on actual behavioural expressions of the Rubondo residents are rather limited, given a paucity of direct observations. Still, over the years, interesting anecdotal or semi-quantitative information emerged which can be related to the paradigm of cultural primatology. As the founders of the Rubondo population were ex-captives, most if not all of these behaviours were likely 'invented' on the island itself.

While some traits represent candidates for arbitrary cultural variants, others represent modes of resource extraction, i.e. carnivory and tool use for extractive foraging, reflecting the maxim 'necessity is the mother of invention'.

Arbitrary Behaviours: Carnivory – Material Culture

Interestingly, the Rubondo apes, seemingly on their own, have independently invented the above described overhead grooming hand-clasp, as observed in Taï, Mahale and Kibale. The existence of the position is evidenced by a photograph of a group of males dated 17Jul19, where two of them groom in the style of the A-frame (Figure 5.12a).

Liza Moscovice describes another seemingly capricious behaviour – not reported elsewhere – which she termed "budding nests" (Moscovice 2006: 192f). In this arrangement, two nests are side-by-side, with some branches from both structures either touching or intertwined, creating the appearance of being connected. Six such budding sleeping platforms were recorded at different spots between Nov02 and Aug03. Typically, one nest was considerably larger than the other. They were mostly built in the same tree, but also found to integrate adjacent canopies. Based on differences in nest size, one occupant was seemingly adult, while the other a sub-adult or juvenile.

However, during our 2012–14 study, we found no evidence for budding nests. Perhaps the trait went unnoticed, because it is rare and exhibited by only one or a few individuals, and not yet disseminated to the larger population. Alternatively, the invention of budding nests might have died out.

Contrary to behaviours associated with resource acquisition, a rapid cycle from local invention to extinction is more likely if the behaviour does not provide much of a tangible benefit. Thus, members of a Costa Rican population of capuchin monkeys would suck each other's toes, stick a finger up each other's nose; or, most strangely, poke a fingertip into the eye-socket of a partner, noticeably dislodging the eye-ball – bizarre fads that somehow developed to then die out after a while (Perry et al. 2003).

Another potential idiosyncrasy of the Rubondo apes was pointed out by Müller & Anzenberger (1995: 66–70). They note, that during 2 months of field research with almost 60 days in the forest, vocalisations were heard only 4 times – of which they were thrice likely provoked by observers getting close to the chimpanzees. This low rate of just 0.07 calls/day contrasts markedly with values of 3.46 calls/day reported for Kibale, Uganda (Ghiglieri 1984) or 4.05 calls/day reported for Gashaka, Nigeria (Sommer et al. 2004). Possibly, the months of Oct–Nov when the 1994 Rubondo study took place are indeed periods of extremely low calling activity, given that at Gashaka there were also months with virtually no calls heard. In any case, it is unlikely that Rubondo fieldworkers did simply not notice emitted calls, given that they carry far and that observers spent much time on hilltops, including during morning and evenings, when elsewhere calls are most frequent. It is also improbable that the absence of conspecific aggression or the low population density causes low call frequencies, as impeding hostile encounters lead to a suppression of call activity while a widely spaced-out community is expected to call more to alert each other about the whereabouts of its members. Instead, Müller & Anzenberger (1995) raise the possibility that the ex-captive origin of the Rubondo apes hampered them – and their descendants – to understand and learn about the functionality of acoustic communication, e.g. with respect to exchanging information. The near-absence of long-distance acoustic ape communication on Rubondo would thus be a reflection of an impoverished cultural repertoire. Perhaps, similar to how nest building skills were only acquired slowly, the apes also developed their acoustic portfolio only over time.

This idea is supported by our very different records for 2012–14, when absence or presence of vocalisations was recorded over 142 days of fieldwork. Calls were heard during all months of the year and during 106 individual days (74.6%), amounting to 557 vocalisation bouts. This translates to 3.92 call bouts per day – a figure strikingly similar to the Kibale and Gashaka data mentioned above. We have no explanation for this considerable discrepancy, except that the 'calling culture' of the Rubondo apes might have changed during the almost 20 years that separate the two reports.

Mwanahamisi Issa Mapua – on Rubondo 2002–07

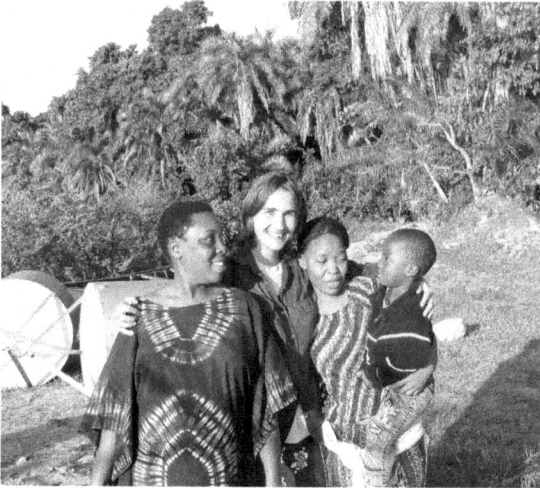

Figure B5.2 Mwanahamisi Issa Mapua (r., holding a child), with field assistant
         Giulia Graziani (m.) and park warden Hobokela Mwamjengwa
         (l.). (Photo: Michael Huffman)

I have always had a keen interest in nature, and after a field trip to
Arusha National Park, I became particularly interested in botany
and ecology. When Liza Moscovice visited Dar es Salaam in 2002,
we made plans that I would support the research that unfolded on
Rubondo. Life in general was quite tough. I was in charge of running
three camps and I rotated between them once per week. I also oversaw
the data collection, ecological monitoring and information gathering
on the chimpanzees. There was no mobile phone reception and com-
municating via walkie talkies was often interrupted as signal strength
varied across the island. Because of its remoteness, we had very few
visitors and did not get a chance to meet different people except our
tracking team and park rangers. It was at times difficult to maintain
morale, so I had to keep the team's spirits high. I also lost count how
many times we were chased by elephants and came close to snake
bites. Often times we got lost in the forest and stumbled to find our
way home to camp. I persevered because to me it was important to
show that a woman can lead a field site. But, looking back, the time
on the island was incredible. At the shore, the water was very clean,
one could take a bath with a beautiful setting sun, birds and otters
popping up and down from the lake waters. The air was so fresh, and
everywhere was so green and full of life.

Carnivory

Across Africa, wild chimpanzees capture and consume small- and medium-sized vertebrates of at least 40 species (review in Newton-Fisher 2015). The most common prey are primates – in particular red colobus monkeys–antelopes and pigs, although mammals such as mongooses, rodents and pangolins are also targeted, along with birds, lizards and frogs. It is mostly males who obtain prey, typically opportunistically, though at some sites active pursuit (hunting) occurs, at times involving several individuals.

The Rubondo inhabitants are known to consume at least two types of mammals, mostly sitatunga antelopes, but also vervet monkeys, as illustrated by the following anecdotal reports:

- sometime 2002–07: "Hunting juvenile sitatungas and vervet monkeys was observed repeatedly" (Petrželková et al. 2010a);
- Oct02–Apr04: "There was one direct observation of chimpanzees killing and eating an immature sitatunga" (Moscovice 2006: 45);
- 13Jul13: Visiting researcher Paco Bertolani and local field assistants surveyed the central Nyakutukula hills. Here, in grassland with few trees, i.e. an area with good visibility, they came across 3 apes. Gradually, numbers built up to 26 individuals, probably because some of them were consuming an infant sitatunga (Chapter 4);
- 15Sep16: Video evidence of a large mammal being eaten (Asilia tourism company);
- 17Aug19: A video recorded by Paul Kivuyo shows a chimpanzee, up in a tree, inspecting the hide of what appears to be an almost entirely consumed sitatunga (Figure 5.12b).

Whenever the age of sitatungas was recorded, the prey was a non-adult (infant, juvenile, immature). We do not know how the antelopes were obtained. However, given that sitatunga calves are left in hiding spots while their mothers forage, chimpanzees travelling on the ground might simply stumble upon them, similar to the opportunistic way baboons catch immature forest antelopes (Sommer et al. 2016). The consumption of sitatungas is not reported from any other chimpanzee habitat.

As for primate prey, at more than a dozen other study sites, chimpanzees consume monkeys, mostly colobines such as red colobus, black and white colobus and olive colobus, but also guenons, baboons and mangabeys (Newton-Fisher 2015, Gilby & Wawrzyniak 2018). To date, we have no indication that Rubondo chimpanzees prey upon the introduced black and white colobus monkeys. One reason might be that these monkeys – while in groups of at least 8–10 – occur mainly near the tourist lodge camp as well as the southern Lukaya ranger station and further south-east in the Msasa hills near Lukukuru (Müller & Anzenberger 1995: 38f), areas rarely frequented by chimpanzees or observers.

Vervet monkeys are eaten at Mahale, Tanzania, but only rarely so, as is the case across Africa – perhaps because these monkeys prefer open woodland, a habitat less used by chimpanzees. While we do not have good descriptions of the reported eating of vervets by the Rubondo apes, they certainly interact with these primates. As a prime example, during a 5-hour episode recorded in 1970 near the guesthouse building, a female chimpanzee was observed to hold and play with a baby vervet (Hans Sönsken, cit. Müller & Anzenberger 1995: 73; cf. Table 2.2).

Given a lack of habituation to observers, our current evidence does not reveal the full picture. In any case, the island harbours numerous large-bodied animal species (cf. Table 3.4), several of which constitute suitable prey. It would not be surprising if more evidence comes to light that some are indeed predated upon.

Figure 5.12a Potential cultural traits of Rubondo chimpanzees. (a) A-frame grooming with overhead hand-clasp. (Photo: JNM) (b) Consumption of a sitatunga antelope; note pattern of hide. (Video-still from Paul Kivuyo) (c) Buttress-beating with rock; note mark on trunk of Ficus tree, and rock lying at base. (Photo from Moscovice 2006: 205) (d) Discarded grass stems used for insect-fishing. (Photo from Petrželková et al. 2010a)

Figure 5.12b Continued

Figure 5.12c Continued

Figure 5.12d Continued

Material Culture

Liza Moscovice describes evidence of a behaviour – again not reported from other study sites – which she termed "buttress-beating with a rock" (Moscovice 2006: 193f). On two occasions, months apart from each other, she recorded markings on the base of a tree trunk consistent with impact from a heavy object. Rocks measuring 15–20 cm x 8–10 cm with pieces of tree bark attached to it were discovered directly next to the buttress root (Figure 5.12c). These finds suggest a scenario in which a chimpanzee pounded a rock against the trunk while sitting or standing. On one occasion, the impacted tree contained fresh nests. During the other occasion, the markings were at the base of a large *Ficus* tree, near where older remains of a deceased subadult or adult male chimpanzee had been found.

The behaviour, not aimed at acquiring a resource such as when nuts are cracked with a stone, has some similarity with a trait termed 'chimpanzee accumulative stone throwing' (Kühl et al. 2016). This refers to a newly discovered lithic tool use reminiscent of human cairns in which chimpanzees – at some locales only – habitually bang and throw rocks against trees, or toss them into tree cavities, resulting in conspicuous assemblages. While on Rubondo, there was no accumulation of stones, rocks were likewise employed in a seemingly 'non-practical' way.

Apart from such 'gratuitous' use of materials, there is some evidence that Rubondo animals also engage in the targeted use of technology. Thus, given that some of their faeces contained termites and ants (see above), this raises the question how the insects were obtained to be ingested. Surely, chimpanzees can catch such prey with bare hands. In fact, shortly after having been released onto Rubondo, a female collected ants from the forest floor to then suck them off from her finger, one by one (Kade 1967). However, chimpanzees elsewhere also insert implements – blades and stems of grass, woody sticks, mid-ribs of leaves – into insect nests, who then, in self-defence, bite into them, allowing the chimpanzees to retract the probe to more easily ferry the prey into the mouth (Fowler et al. 2011).

Liza Moscovice, summing up her 16-months study from 2002–04, states a "lack of direct or indirect evidence for material culture" associated with extractive foraging (Moscovice 2006: 191). Indeed, for Rubondo, till this date, evidence of material culture involving the foraging for social insects is all but absent. As an exception, sometime during 2004–07, discarded fresh grass stems were found near what seemed like entrance holes to an underground nest of insects (Figure 5.12d). While the observers (Petrželková et al. 2010b) assume that these were ants, they were more likely termites of the genus *Macrotermes* or *Pseudacanthotermes* (Alejandra Pascual-Garrido, pers. comm.). Similarly, during our 2012–14 research, modified 'probing sticks' seemingly fit for insect extraction were only found once, in Jul13 in the Kasenya sector.

Michael Huffman (pers. comm.) kindly shared information about what he considers another example of tool use. Thus, in May05, on a visit to Rubondo, he and Mwanahamisi Issa Mapua found what they thought was a tool set, potentially used by chimpanzees for extracting termites from a mound at the base of a tree. Scattered around were several well-worn tools, consisting of modified branches. A single thick sturdy branch might have been employed to open semi-closed termite mound entrances, while multiple probing sticks extracted the insects from inside. Upon inserting such stick, several termites clung onto it and were fished out quite easily. Only one such mound was found, but it appeared to be frequently used. The observers asked the accompanying trackers if humans could have done this, and they responded that they had never heard anybody talk about termite fishing. Thus, the possibility remains that chimpanzees had indeed extracted the termites, albeit not by using strips of bark or grass as is typical for East African apes, but with a tool set similar to what has been documented in Central Africa (cf. Almeida-Warren et al. 2017). Another expert (Alejandra Pascual-Garrido, pers. comm.) believes the targeted termite taxon is likely *Pseudacanthotermes*, which reproduces around April, while disagreeing with the idea that chimpanzees were involved, instead finding it more probable that humans fashioned the tools, aiming to provoke alates to the surface to then snack on them.

Taken together, these anecdotal observations constitute – indirect – evidence that the descendants of introduced chimpanzees are capable of using tools. Elsewhere, released ex-captives have likewise developed functional responses, with respect to both prey capture (eating of monkeys) as well material culture (nut-cracking and ant-extraction with tools) – patterns we will describe in more detail when reviewing other translocations of chimpanzees (Chapter 6).

Still, all said and done, the Rubondo apes employ material culture with extremely low frequency. This suggests that such rare actions "may not be habitual or customary traditions" but rather "novel behaviours exhibited by only one or a few individuals, which have not disseminated to the larger population" (Moscovice 2006: 194). One causal factor could be the low population density. If individuals are dispersed across a large range, this limits the opportunities for innovative behaviours to spread once they have sprung up. In addition, cumulative technical progress and innovation develop more likely when ecological and social conditions are unstable, with environmental variability spurring a richer behavioural diversity (Kühl et al. 2019).

There are other populations, where tool use is rare resp. absent, e.g. at Budongo, Uganda and at bonobo sites such as LuiKotale, DR Congo (McGrew et al. 2007). At both locales, there is an abundance of edible plants, i.e. fruit resp. terrestrial herbaceous vegetation. In the same vein, at Rubondo, high-quality fleshy fruit is available across seasons. Conversely, a more challenging habitat such as Gashaka, Nigeria, is associated with an extensive tool kit (Sommer et al. 2011). The scarcity of material culture or its infrequent practice, rather than implying behavioural limitations of ex-captives, therefore probably reflects relaxed ecological constraints (Moscovice 2006: 191) – as opposed to the mentioned principle of 'necessity is the mother of invention'.

Nevertheless, we might rightly be curious about yet-to-be-discovered behaviours once the chimpanzees allow the close presence of inquisitive human observers.

6 Apes in the Anthropocene: Lessons from a Maverick Release?

Figure 6.1 Rubondo chimpanzee. (Photo: Rob Slater, Safari Consultants)

Half a century on, with the benefit of hindsight, how are we to view and judge those historic chimpanzee translocations from Europe to Lake Victoria?

Initially, the impetus had been to turn the isle into an "artificial game reserve" (Peter Achard). Yet, the Rubondo releases, by those who initiated and executed them, have been called a "home-coming" (Sinclair Dunnett) and a "Wiedereinbürgerung" (reintroduction, Berhard Grzimek). These conflicting descriptors (Chapter 1) raise obvious questions. Can an isolated chunk of biome that was substantially redesigned by human intervention still pass as a natural habitat? Can the translocation of apes from different subspecies to a region outside their native ranges pass as a reintroduction (Figure 6.1)? Does liberating some apes have any impact on species

DOI: 10.4324/9780367822781-6

conservation? Were the releases a success or a failure? Can they teach us something in terms of conservation politics or welfare policies? And finally, if we could travel back in time and do it all over again, is there anything we would do differently?

To reflect on these questions, we review outcomes against intentions – and alongside similar efforts in other parts of Africa.

Uprooting Apes: An Age-old Trade

Today, the principal threats to the survival of wild primates including non-human apes are habitat destruction and hunting (Peterson & Ammann 2003). However, historically, the animals themselves were viewed as valuable and desirable items (for the following, see Stiles et al. 2013) – a fate shared by the future Rubondo founders.

Already the biblical Old Testament mentions that King Solomon "had a fleet of trading ships" which "returned, carrying gold, silver and ivory, and apes and baboons" (1 Kings 10:22). The trading of apes as a commodity became increasingly global from the 15th century onwards, when Europeans colonized distant parts of the globe and explorers brought back from their voyages not only precious minerals and spices, but also exotic plants and animals. The imported creatures were often displayed in royal menageries, signifying that a ruler's influence was far and wide. Soon, well-to do citizens took up the hobby to showcase in their home non-native species for family entertainment and to impress visitors. In the early 19th century, many European cities built and opened public zoos, while travelling circuses became popular, too. A hundred years or so later, when their close genetic ties to humans had been established, chimpanzees were increasingly sought after as subjects for biomedical experiments, scientific endeavours such as space programmes, as well as for behavioural and cognitive research, especially in the USA (Yerkes & Learned 1925, Kellogg & Kellogg 1933) and Russia (Ladygina-Kohts 2002).

The full extent of the associated trade is not known. Chimpanzees were shipped mainly from West Africa and in particular Sierra Leone, from where, between 1950 and 1980, about 2,000 were exported to biomedical labs, zoos and circuses, and as pets (Teleki 1989, Peterson & Goodall 1993). Pointing out similarities between the violence humans have wrought against other humans and the treatment of non-human animals is a sensitive political issue. Still, one is tempted to make that "dreaded comparison" (Spiegel 1988) between the trade in non-human apes and the extraction of millions of human slaves from this region, another 'commodity' prized by North Atlantic nations (Harris & Lusted 2019). The 2014 'European Studbook for *Pan troglodytes*' (Carlsen & de Jongh 2014) contains historical information on 3,906 chimpanzees (46% males, 54% females), of which 42.0% were wild-born. The first record dates back to 1835, and the last – of a wild caught ape from Nigeria – to 1999. As we will detail later on, one zoo resident is equivalent to at least ten apes that are killed to get hold of this single individual. Therefore, European zoos alone are responsible for the fate of 18,084 apes, with a minimum of 16,440 slaughtered in the wild, to obtain 1,644 destined for captivity.

Apes have not always been legally protected. In fact, only recently were laws passed and agreements signed to halt the hunting, killing and trade in great apes – notably the Convention on International Trade in Endangered Species of Wild Fauna and Flora (CITES) from 1973, the Endangered Species Act in the USA from 1973, and the African Convention on the Conservation of Nature and Natural Resources from 1968. The latter designated apes as a 'Class A' species, which "shall be totally protected throughout the entire territory of the Contracting States".

Consequently, by the mid-1970s, fewer wild-born chimpanzees were exported to Europe and America, and because, by the 1980s, survival rates in captivity had markedly improved, this further reduced the need to acquire specimens from habitat countries. At the same time, various breeding colonies were set up to supply biomedical facilities. However, new markets soon opened up in other parts of the globe. So, just as the European thirst for wild apes quelled, it grew in Asia and the Gulf States. Today, with shifted commercial epicentres, apes and other primates are still a sought-after good, trafficked in an organised and criminal fashion, to serve as status-enhancing pets of wealthy owners or in the entertainment industry (Stiles et al. 2013, Nekaris & Bergin 2017).

The targeted extraction of wild animals does not only feed supplies to menageries, zoos, labs or pet-owners. Apes in particular are also prized as dead bodies – as trophies, ingredients for traditional medicine and as edible meat (Nyanganji et al. 2010). Book titles, such as *Consuming Nature* (Rose et al. 2003), *Slaughter of the Apes* (Ammann & Pearce 1995) or *Eating Apes* (Peterson & Ammann 2003) reflect upon these practices (see also Ammann & Sommer 1998).

Subsistence hunting has been taking place since the dawn of humankind and meat from wild animals is a conventional source of protein across tropical Africa. Yet, since about the 1980s, the bushmeat trade evolved into a fully fledged commercial business, highly profitable with a turnover estimated at billions of dollars (Fa et al. 2003). In the Congo Basin alone, between one and five million metric tons are eaten each year. The most commonly consumed species are ungulates and rodents. Primates constitute up to 40% of the species on offer at a typical bushmeat market, with great apes making up 1–2.5%. Much of the demand comes from expanding urban centres, facilitated by the opening of previously inaccessible areas by logging and mining companies. The easy availability of guns in the wake of armed conflicts (Benz & Benz-Schwarzburg 2010) has had a devastating impact on larger bodied animals in particular, as heavier specimens are preferentially targeted to get more profit per cartridge.

In addition to apes being directly targeted, they are also increasingly marginalised by the needs of an ever-growing human population that has caused the rampant conversion of natural forests (Arcus 2015). This often leaves primates, including chimpanzees, in isolated, small pockets which are even more vulnerable because of edge effects and associated dangers of the spread of diseases (Hill et al. 2001).

Today, as a result of dramatic decline in numbers – around 50% in the last 75 years – *Pan troglodytes* is listed as 'endangered' by CITES. All range states have signed up to the CITES treaty and additional national and international laws make it illegal to kill, capture or trade in live animals or their body parts. However, such stipulations often exist on paper only and are not or very rarely enforced on the ground.

How, in the face of unabated trade, habitat loss and disease transmission, are chimpanzees fairing today? One estimate, however error-prone to arrive at, puts numbers remaining in the wild at 345,000–470,000 (Humle et al. 2016). *P. t. ellioti* is the most endangered subspecies (6,000–9,000 individuals remaining), followed by *P. t. verus* (18,000–65,000), while larger numbers remain of *P. t. troglodytes* (140,000) and *P. t. schweinfurthii* (181,000–256,000). Some countries have lost up to 90% of their chimpanzees in as little as 20 years (Campbell et al. 2008), a trend that is expected to continue. We may well be the last generation of humans that coexists with viable populations of wild chimpanzees.

Apes in Sanctuaries: The Tip of an Iceberg

From the late 1970s onwards, as legal changes rendered it more difficult to traffic wild caught primates and exploit them as pets or for entertainment purposes, many illegally kept apes were confiscated. These orphans needed caretakers. With hardly any institutionalised solutions in place, some well-meaning people began to house stranded apes – not because they wanted to be pet owners, but because they were moved by the plight of these human-like creatures. Over the years, such private initiatives, often involving ex-pats, led to the formation of official 'sanctuaries' and associated non-profit organisations and NGOs to support the centres (cf. narratives in Brewer 1978, Carter 1981).

At present, there are about 20 ape sanctuaries in 12 African countries, and 18 of these house chimpanzees – currently about 1,200 (Table 6.1). Almost all of these refuges are organised under the umbrella of the Pan African Sanctuary Alliance (PASA). Established in 2000, PASA improves communication between managers, develops standards for husbandry, health and reproduction and advocates nature conservation (Farmer 2002b, Faust et al. 2011).

One of the first cross-sanctuary censuses was led by PASA co-founder Norm Rosen, in conjunction with this book's co-author (VS) and researchers Nicola Hughes and Neil Gretsky (Hughes et al. 2011). This account

Table 6.1 Chimpanzees kept in sanctuaries across Africa (2017)

Country	Sanctuary name	Founded	Population (2017)
Cameroon	Ape Action Africa	1996	110
	Limbe Wildlife Centre	1993	55
	Sanaga–Yong Chimpanzee Rescue	1999	72
Congo	HELP–Congo	1990	30
	Tchimpounga Chimpanzee Rehabilitation Centre	1992	160
D.R. Congo	Centre de Rehabilitation des Primates de Lwiro	2002	72
	Jeunes Animaux Confisques au Katanga	2006	38
Gambia	Chimpanzee Rehabilitation Project	2008	99
Guinea	Centre pour Conservation des Chimpanzees	1997	60
Kenya	Sweetwaters Chimpanzee Sanctuary	1993	36
Nigeria	Afi Mountain Wildlife Sanctuary	2000	43
	Pandrillus	1991	31
Sierra Leone	Tacugama Chimpanzee Sanctuary	1995	100
South Africa	Chimpanzee Eden	2006	32
Uganda	Ngamba Island Chimpanzee Sanctuary	1997	49
Zambia	Chimfunshi Wildlife Orphanage	1988	132
Liberia	Vilab project	2006	66
Total number of chimpanzees			1185

Source: centre websites Sep17

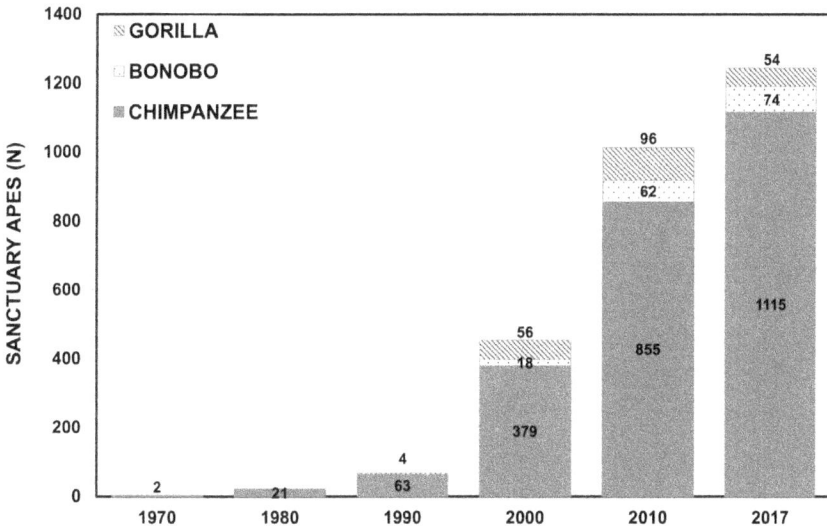

Figure 6.2 Development of great ape numbers kept in sanctuaries across Africa. (After Hughes et al. 2011; PASA 2019)

revealed the staggering development with which chimpanzees, gorillas, and bonobos have ended up in institutionalised care (Figure 6.2). At first, only a few animals were kept, and growth was slow, from 2 (reference year 1970) to 21 (1980) and then 63 (1990). In the 1990s intake exploded, first to 379 (2000), thereafter to 855 (2010) and then to 1,115 (approaching 2020). Overall, the intake was 1.9 animals per year until 1980; it then jumped to 17.9 animals for the subsequent years till 2000, where it peaked at 39.8 animals per annum for the years approaching 2020.

Intakes for bonobos and gorillas followed a similar trajectory, albeit these taxa make up only about 16% of sanctuary apes. Their much smaller geographical range and ongoing conflict in their home nation, DR Congo, explains why there are far fewer bonobos than chimpanzees in sanctuaries. Although gorillas have a larger geographical distribution, their infants, once ripped from the natal group, are very delicate, succumbing easily to captive conditions – which explains their low sanctuary presence.

Of course, the relatively lucky denizens of such oases embody only a fraction of those that were killed in the process of capturing a live infant. To understand the lethal dimensions behind the sanctuary populace, contexts and demographic features were analysed in more detail for chimpanzees housed in seven sanctuaries in Cameroon (C) and Nigeria (N): Sanaga-Yong Chimpanzee Rescue Centre (C); CWAF (Cameroon Wildlife Aid Fund) (C); Mvog-Betsi Zoo Yaoundé (C); CWAF Mefou National Park (C); Limbe Wildlife Centre (C); DRBC (Pandrillus/Drill Rehabilitation and Breeding Centre), Calabar (N); DRBC, Afi Mountain Wildlife Reserve (N). Data about each of 161 arrivals that had passed through the facilities between 1986 and 2003 were gathered from written records and interviews with personnel (Hughes et al. 2011) – no mean task, as one of the last things on the minds of overworked caretakers is the keeping of retrievable records (Figure 6.3).

The sex ratio of arrivals was relatively even, with 53% males and 47% females. The vast majority (80.1%) were younger than 4 years (23.6% <1 year; 31.3% 1–2 years; 25.5% 3–4 years) and thus removed from the wild as dependent, nursing infants. About half (50.5%) of all occupants were donated, while 43.0% were confiscated. Relatively few animals (6.5%) arrived by other means (abandoned, zoo transfer, unknown). Before their arrival, about one-third (30.4%) were kept as pets in homes, one-fifth (21.2%) held in villages, and another fifth (22.4%) had been on show in restaurants, bars, hotels, zoos or amusement parks.

Due to their slow reproductive rates, even low-level hunting severely threatens ape populations. However, ascertaining the scale of the killings is a notoriously difficult task. Market surveys will likely lead to underestimations, because carcasses will be chopped up and smoked and ape meat will often not be on display, as hunting and sale are illegal. Also, the meat often

Figure 6.3 Caretaker feeding orphaned chimpanzees at Tacugama Sanctuary in Sierra Leone. (Photo: Kat Brinsley)

bypasses open markets and is directly sold to well-to-do buyers, including the African diaspora (Wood et al. 2014).

Sanctuary intake rates offer an alternative window into the dimensions of the bushmeat trade. An important starting point in this modelling attempt is the realisation that, because infants are so small, hunters make more profit by selling them alive. Yet, infant apes are not specifically targeted to extract them from the wild. Instead, they typically end up in human abodes as by-products or bonuses of meat hunting. During many years, when Swiss wildlife photographer and conservationist Karl Ammann investigated the bushmeat trade, he never came across a case where direct demand for a live infant had prompted the extermination of an ape group (Ammann & Pearce 1995; Peterson & Ammann 2003). Instead, some infants simply happen to survive the indiscriminate slaughter of a hunt. A typical pursuit begins with poachers listening for the calls of chimpanzees about to build their sleeping platforms. They then approach the resting site before dawn, waiting until there is sufficient light. Hunters cannot normally pinpoint trees or nests that shelter a mother and her infant. Instead, they fire indiscriminately at any ape they happen to see, while the group scrambles to escape. Sometimes, infants, being shielded by their mother's belly, survive the random shooting.

These practices and the fact that 80% of all sanctuary chimpanzees were taken as infants from the wild allow us to employ sanctuary intake rates as corollaries of hunting pressure.

How can we approximate the number of dead chimpanzees a sanctuary orphan represents? Because infants cannot be captured other than by maiming or killing their mothers, each orphan represents at least 2 individuals lost to the wild. But, given that a targeted shooting of just a nursing mother is not realistic, other group members must be victims, too. Up to 4–5 years, chimpanzees are classified as 'infants', given that they nurse, ride on the back of conspecifics and seek constant association with their mothers. The proportion of infants in well-studied communities across Africa averages 23.6% (11.2%–38.1%; figure based on 66 census years encompassing 3,661 chimpanzees including 842 infants; Hughes et al. 2011). Correspondingly, there are 76.4% non-infants in a hypothetical community. Consequently, (76.4/23.6 =) 3.2 additional individuals have to be killed to obtain an infant by chance. However, this is an underestimate, as mother–infant pairs often nest alone and are therefore less often targeted than larger, noisier nesting parties. This again increases the proportion of wild comrades killed, until, statistically, a single infant is caught. In addition, during a hunting massacre, many infants are shot outright along with other group members, receive injuries that prove fatal or perish on the way to being sold and while being held captive by people. Of the survivors, few actually make it to a sanctuary, because authorities are unwilling to confiscate, or because refuges are full. Given such lines of thought, the Jane Goodall Institute estimates that one sanctuary chimpanzee equates 10 removed from the wild – the surviving infant and 9 others who were killed (Peterson & Goodall 1993, cf. also Teleki 1989). In all likelihood, even this calculation is an underestimate, and a single orphan that somehow made it into a sanctuary may well represent 20, 30 or 50 conspecifics who perished. The current sanctuary population of ca. 1,200 individuals may therefore easily point to more than 50,000 chimpanzees recently lost to the bushmeat trade. After all, by its most optimistic estimate, less than 500,000 chimpanzees survive in the wild, while in the not-too-distant past, there were many millions. These animals did not simply disappear into thin air or die peacefully.

The vast majority of the surviving 'refugees' (Teleki 1989, Palmer 2020: 64f) are deeply traumatised. Ripped from their mothers during the brutal and deadly crackdown on their natal group, they spend considerable time in miserably captivity, often chained around neck or ankles – not to mention bad food, exposure to disease and social deprivation. Accordingly, sanctuaries take on considerable responsibility when accepting a dependant, as it comes with the commitment of lifelong care for a long-lived animal. This difficult long-term prospect and financial burden, as well as the assumption that an independent life in the forest might be a better choice, has led sanctuary managers to consider if and how they can provide a more independent life for their charges – ideally, by returning them back to the wild.

Primate Translocations: Historical and Terminological Perspectives

In the following, we prepare the ground for a comparison of the Rubondo operations with similar releases – as it is through such juxtaposition that we can put their relative success or failure in perspective.

First, we need to detail the terminological framework and important examples of animal translocations. Ape releases are part of broader procedures, whereby humans transport organisms to new regions outside their range, a mode of dispersal called *ethnophoresy* (Heinsohn 2003). The resulting artificial populations come in various shapes and sizes. Many were set up by traders and colonists for commercial, recreational or aesthetic reasons, or as semi-free-ranging groups for biomedical or behavioural studies. Additionally, animals at times escaped from research facilities, zoos or visitor attractions.

For example, to satisfy carnivorous human cravings, Nile perch and Tilapia fish were brought to Lake Victoria (Chapter 3), rabbits to Australia, sheep to New Zealand and West African monkeys to Caribbean islands. Monkeys were also taken from India in 1938 to establish a breeding colony of rhesus macaques for behavioural and biomedical research on the islet of Cayo Santiago off the coast of Puerto Rico – still in existence today (Abee et al. 2012). At times, so-called alien species end up in new places more or less accidentally. Thus, Burmese pythons stemming from unwanted or escaped pets are now thriving in south Florida (Meshaka et al. 2004) and hippopotamuses roam in Columbia, a legacy from the private zoo of drugs-trafficker Pablo Escobar (Kremer 2014). In India, rhesus monkeys perceived as pests by locals are at times moved to other places, including bonnet macaque habitat, which has triggered a decline of the endemics (Radhakrishna 2017). Within this long history of human-aided animal dispersals, it is not until the 1960s that they begin to be tied to conservation and welfare issues (Seddon et al. 2007), aiming to preserve threatened species or as an alternative to a life behind bars (Dore 2017; see also Grzimek 1966c).

The multitude of reasons and modes why and how animals are moved from one place to another does not lend itself to a standardised overarching terminology (Ewen et al. 2013). Our system (adapted from Beck 2018: 1–10, see below) aims to keep definitions simple, albeit we are aware that all categories are prone to ambiguity.

The broadest umbrella-term is *translocation*, which specifies events in which humans move animals from one place (*origin* site) to another (*destination* site). The term *capture* applies when free-living animals are translocated into a captive setting, whereas *release* describes a liberation from captivity. The freed animals are termed *releasees*. Their freedom comes in degrees, as animals may be let out into a wild, unbound habitat, onto islands where movement is confined by water, or into large enclosures with naturally growing vegetation. To qualify as a release, the animals must have "constant interaction with other native fauna and flora" (ibid., p. 3), i.e. be part of the biodiversity of that habitat.

Translocations can be further specified according to their *purpose*:

- *Conservation* releases (also described as *rewilding* or *restocking*) include (i) *reinforcement* of native populations, (ii) *reintroduction* into habitats where the natives have disappeared, (iii) *introduction* outside the historic range (also called *assisted colonisation, assisted migration,* or *managed relocation*).
- *Welfare* releases include (i) *rehabilitation* of ex-captives from zoos, sanctuaries or research centers, (ii) *control*, i.e. moving 'problem' animals, such as those unwanted by owners or those who inflicted damage to people, habitations or cultivations, (iii) *rescue*, i.e. relocating animals endangered by anthropogenic activities, such as dam construction, (iv) *re-unification*, i.e. sending animals from temporary human care back into their natal groups.
- *Commercial* releases are intended to start breeding colonies for biomedical purposes or to attract visitors to the site.
- *Scientific* releases typically attempt to document how translocated animals fared in order to improve survival outcomes; associated strategies and technologies may aim to monitor the animals post release, e.g. by using telemetry.
- *Other* releases encompass those for (i) *touristic* (aesthetic, recreational, religious) reasons, as well as (ii) accidental *escapes*.

A given translocation can embody several purposes. For example, an animal kept in a small holding cage in a sanctuary might be radio-collared and set free in a surrounding national park, which would serve welfare, scientific and touristic purposes.

With a specific focus on primates, Benjamin Beck discusses the ins and outs of these definitions in his comprehensive account *Unwitting Travelers: A History of Primate Reintroductions* (Beck 2018). (Note that he calls every translocation a 'reintroduction', different from the system we are using.) About the animals concerned, Beck writes: "We took them intentionally, enabled by technology and human agency, fueled by human arrogance, and replete with the full range of noble to crassly self-serving motivations with which we treat our fellow primates" (ibid., p. i). As the primates themselves "did not consent to the journeys and had no idea where they were going or why", he calls them "unwitting travelers". Beck is the foremost authority on primate translocations. As a comparative psychologist specialising in animal cognition, biodiversity conservation, he was primate curator at Brookfield Zoo (USA) from 1970 to 1982, and coordinated the reintroduction and post-release monitoring of 146 golden lion tamarins in Brazil between 1983 and 2005. They have multiplied to 1,000 animals by now. Given his experience, Beck is also the lead author of the *Best Practice Guidelines for the Re-introduction of Great Apes* (Beck et al. 2007), which we will discuss later.

In his book, Beck reviews 234 translocation events of apes, old world monkeys, new world monkeys, lorises and lemurs, that involved 24,212 individuals (Beck 2018: iii). Of these, 95.9% were wild-born and 4.1% captive-born. There were 329 identifiable purposes, most commonly welfare (46%), followed by research (20%), conservation (15%), commercial (12%) and other (7%).

Whether or not the moving of animals is deemed 'successful', depends on the definition. According to Beck (2018: v), a lower-bar success means "survival of some of the released individuals for at least a year, post-release reproduction and survival without human support", whereas a higher-bar success requires "contribution to the establishment of a self-sustaining wild population". The mentioned repopulation of ancient habitat in Brazil with golden lion tamarins constitutes a rare gold-standard here (Kierulff et al. 2012).

Note that the higher-bar measure is solely framed in terms of conservation biology; by this yardstick, successful releases are only those that prop up dwindling wild populations or replace already extinct ones. Welfare aspects find no consideration, for example, when freed primates are assimilated by a viable wild population or set free on isolated islands. We will revisit the distinction – or tension – between conservation- and welfare-related releases later on.

Chimpanzee Releases: A Jumble of Approaches and Outcomes

The first documented release of any type of great ape took place in 1962 in Malaysian Borneo, when three orangutans were moved to Bako National Park (Harrison 1963 cit. in Beck 2018: 86–88). As for African apes – gorilla, chimpanzee, bonobo – the earliest release is indeed the one on Rubondo from 1966 to 1969.

We now take the reader through the entire catalogue of 21 recorded cases when chimpanzees were set free on islands or in mainland forests, including some instances of accidental escapes and reuniting infants with their natal groups. While we do not cite sources in our brief narratives, a selection is listed in the summary overview (Table 6.2). All but one of the events are covered in Beck's (2018) already mentioned comprehensive compilation about primate reintroductions. Other important cross-site accounts are found in Hannah and McGrew (1991), Teleki (2001), Farmer (2002a), Carter (2003) and Beck (2010).

Table 6.2 Chimpanzee translocations into the wild or onto islands 1966–2015

Case	Site	Country	Date	Purpose	Number, origin	M:F cohorts	Age (yrs)	Rehabilitation	Provisioning	Context, problems	Outcome	Source
1	Rubondo Il NP, Lake Victoria (237 km²)	Tanzania	1966–69	1-3, 2-2, 5-1	16 (W) in 4 cohorts	3:7, 1:0, 1:0, 2:2	4–12	N	N (Y = short-term)	1 F died in transit. People bitten by several animals. 1–3 aggressive M shot by rangers	Reproducing population, wild foraging, antelope hunting, likely tool use; ~35 by 2014	Msindai et al. 2021, this book
2	Ipassa II, Ivindo River (0.7 km²)	Gabon	1968–72	2-2, 5-2	8 (W)	3:5	4–8	N	Y (30% of diet)	1978: some individuals waded to mainland	Ate small mammals and birds. Most recaptured and returned to laboratory; some remained on mainland	Hladik 1973, 1977, Beck 2018: 21
3	Bear Il YRPC (0.01 km²)	USA	1972–73	3	8 (5W, 3C)	1:7	Adult	Socialisation	Y	1 F died of pneumonia, 1 F died after miscarriage, 1 F failed to nurse offspring. Released into temperate climate, but cages not heated	Built nests in trees, ate wild foods. Plans for biomedical research colony abandoned in 1974; 6 remaining returned to primate center	McGrew 1983, Beck 2018: 41–42
4	Mt. Assirik, Niokolo Koba NP	Senegal	1972–79	2-1	14 (12W, 2C) in 8 cohorts	9:5	2–10	Y	Y	3 young died in transit; 4–5 disappeared post release; several attacked by resident wild chimpanzees	Remaining 7 plus 1 infant transferred to River Gambia (Case 5)	Brewer 1978, 1980, Brewer et al. 2006, Beck 2018: 23–25
5	Baboon Il (Island 2), River Gambia NP (0.8 km²)	Gambia	1979–85	2-1	8 (5W, 3C), 16	?, 7:10	1–13	Y	Y	2 F disappeared	Moved in 1985 to larger Baboon Il (Case 7)	Brewer 1980, Carter 2003, Brewer et al. 2006, Beck 2018: 26–27

#	Site	Country	Year									References
6	Baboon IL. (Island 1), River Gambia NP (4.4 km²)	Gambia	1979	2-1	9 (7W, 2C)	3:6	3–11	Y	Y	1 died of endoparasitism. Captive-born Fs adopted wild-born orphans	1 F captured and ate a monkey; 1 infant born; merged in 1985 with above group (Case 7)	Carter 2003, Brewer et al. 2006, Beck 2018: 27–29
7	Baboon IL. (Islands 1, 2, 3), River Gambia NP (4.4 km², 0.8 km², 0.5 km²)	Gambia	1985–2003	2-1, 5-1	8 (Case 6), 14 (Case 5), 33 (W)		Infant to adult	Y	Y	1 F killed by local people in 1987; some mortality; continued provisioning necessary; likely mixed subspecies composition; no contraception	By 2002 59 individuals (33:26) with 39 born on islands, by 2016 116 individuals; of 145 newborns, 70% were still alive	Carter 2003, Brewer et al. 2006, Beck 2018: 30–34
8	Little Bassa River (5–6 IL) (0.1–0.4 km²)	Liberia	1978–2006	2-2	125 (W) in 6 cohorts	?; ?; 10:12, 8:10, 19:11	5–20+	Y	Y	ViLab colony. Many died, excacerbated by civil war; some drowned; conspecific aggression with at least 1 M killed; disease, starvation; attacks on humans, with at least 1 serious injury	While on the islands, many chimpanzees ate wild foods, used hammer-stones, built nests; some newborns; by 2016, 66 still alive (~50%)	Hannah & McGrew 1991, Carter 2003, Beck 2018: 34–39
9	Bandama River IL, Azagny NP (1.7 km²)	Ivory Coast	1983	2-1, 2-2	20 (W)	10:10	7–11	?	Y	ViLab colony. Many died from untreated suspected Shigella; by end of 1984, only 2 remained; by 2013, only 1 M alive	Since 2003, a local villager supports the survivor(s)	Teleki 2001, Chonghaile 2002, Carter 2002, 2003, Clifton 2015, Beck 2018: 39–41
10	Kerfalya IL, Konkoure River (0.01 km²)	Guinea	1992	2-1	12 (W)	6:6		?	Y	10 died within 2 yrs, remaining 2 had died by 2000		Carter 2003, Beck 2018: 47f

(Continued)

Table 6.2 Continued

Case	Site	Country	Date	Purpose	Number, origin	M:F cohorts	Age (yrs)	Rehabi-litation	Provi-sioning	Context, problems	Outcome	Source
11	Kibale NP (770 km²)	Uganda	1994	2-1	1 (W)	0:1	5	3 wk	N	Attempt to release confiscated pet; showed up after 10 days in village searching for food; returned to wild chimpanzees subsequently again found in village	Released into Kanyawara study community; refused to go back to forest after 1.5 mo; taken to Entebbe zoo, then Ngamba IL	Treves & Naughton-Treves 1997, Beck 2018: 13–14
12	Isinga IL, Lake Edward (0.1 km²)	Uganda	1995–96	2-1	12 (W)	5:7	Infant to adult	?	Y	3 from European zoos, plus confiscated orphans. After several days, 2 infants found dead in water	Plans for later release into larger area abandoned in 1998; 10 survivors moved to Ngamba IL (Case 13)	Manning 1996, Beck 2018: 17–19
13	Ngamba IL SA, Lake Victoria (0.4 km²)	Uganda	1998	2-1, 5-1	15 (W)	2:8	Infant to adult	Y	Y	Enticed to enter holding facility for sleeping each night	Provisioned; tourism supports facility; 49 chimpanzees in 2018, due to additional transfers and some births	Cox 1998, Beck 2018: 19f
14	Afi Mountain Wildlife SA (104 km²)	Nigeria	2000	5-2	4 (W)	3:1	5–12	N	N	Accidental escape; no crop-raiding, no conflict with humans	1 recaptured next day, 1 more 3 mo later; 2 survive in the wild for at least 3 mo; 1 possibly killed by hunters	Farmer 2002, Carter 2003, Beck 2018: 48f

15	Conkouati-Douli NP (5,050 km²)	Congo	1996–2001, 2005–12	2-1, 1-1, 4	37 (W), 16	10:27; ?	Infant to adult	Y	9 M, 9 F attacked by wild conspecifics, ~6 died; higher mortality mitigated by veterinarians	1996–2001: several joined wild groups; 2005–2012: scant information; by 2012, 17 alive, 20 disappeared, 15 dead, 1 returned	N (Y = minimal)	Beck et al. 1994, Farmer 2002, Goossens et al. 2002, Beck 2018: 42–47
16	Tacugama Chimpanzee Sanctuary, Western Area Peninsula NP (177 km²)	Sierra Leone	1998, 2000	5-2	5 (W), 31 (27W, 4C)	1:3; 16:15	3–20	Y	Late 1990s: accidental escape during forest walk; 2006: escape after juvenile opened enclosure; 1 human killed, another severely injured	Late 1990s: 3 recaptured within 1 yr, all in good health; 1 seen 1 yr post escape. 2006: 21 returned voluntarily, 6 recaptured, 4 evaded	N	Farmer 2002, Carter 2003, Kabasawa et al. 2008, Beck 2018: 49f
17	Diarra River	Senegal	2002	2-4	1		1.5–1.8	N	Villager found infant next to dead F	Released into natal group after 1 mo in captivity; seen several times thereafter	Y (short-term)	Beck 2018: 51
18	Unkown	Guinea	Before 2013	2-4	2		Infant		Orphaned infants, in captivity <6 mo	Released near sites of their capture		Beck 2018: 51
19	Haut Niger NP (554 km²)	Guinea	2008, 2011	1-1, 2-1	12 (10W, 2C), 4 (W)	6:6; 2:2	8–20	Y	1 M died upon recapture, 1 M injured by wild chimpanzees, 3 disappeared; 1 F (in company of 1 wild F) returned to sanctuary	2008 cohort: 6 of 12 releasees survive till 2009, 1 F joined wild group; 2011 cohort: scant information; several newborn	N (Y = short-term)	Humle et al. 2011, 2013, Beck 2018: 52–54

(*Continued*)

Table 6.2 Continued

Case	Site	Country	Date	Purpose	Number, origin	M:F cohorts	Age (yrs)	Rehabilitation	Provisioning	Context, problems	Outcome	Source
20	Okong IL, Pongo IL, Sanaga River, Douala-Edea Wildlife RS (0.9 km², 0.5 km²)	Cameroon	2008, 2010, 2015	2-1, 5-1	9 (W), 6 (W), 1 (W)	6:3; 4:2; 0:1	9–16	Y	Y	1 M killed by local villagers, all others survived	Hunting of monkeys; 5 newborns from 3 F; by 2017, 20 animals total	Angwafo et al. 2018
21	Fongoli Savanna Chimpanzee Project site	Senegal	2009	2-4	1 (W)		9 mo	N	Y (1 wk)	1 F of study group attacked by hunters and her infant taken	Released into natal group after 1 wk in captivity; both mother and infant survived	Pruetz & Kante 2010, Beck 2018: 54–56

Site: IL = island, NP = national park, RS = reserve, SA = sanctuary; *release purpose:* 1 = conservation, 1-1 = reinforcement, 1-2 = reintroduction, 1-3 = introduction, 2 = welfare, 2-1 = rehabilitation, 2-2 = control, 2-3 = rescue, 2-4 = reunification, 3 = commercial, 4 = scientific, 5 = other, 5-1 = touristic, 5-2 = escapes; *origin:* C = captive-born, W = wild-born; *sex:* F = female, M = male; *rehabilitation/provisioning:* N = yes, Y = yes; *source:* in a few instances, information from other reports was prioritised over Beck (2018). (Date of reference: 2018)

We present the releases chronologically, starting with the first-ever operations, those on Rubondo – which, for context, we summarise in the style of the subsequent reviews.

- Case 1: Rubondo Island, Tanzania (1966–69). Over the course of three years, 17 ex-captive chimpanzees born wild in West or Central Africa were transported from Europe to East Africa. One animal died in transit, so, 16 were set free. There had been no attempts to familiarise the animals with a life in the wild. Still, they survived on wild foods, built nests in trees and began to reproduce. Several people were bitten by the apes. At least 2 aggressive males were shot by rangers. Given sketchy post-release monitoring, it remains unknown if other animals died. Survival on the large island was aided by abundant food sources, absence of other apes or large terrestrial predators and effective habitat protection. Today, the island harbours 35 or so chimpanzees.
- Case 2: Ipassa Island, Gabon (1968–72). Stemming from a medical-research facility, the 8 released apes built sleeping platforms in trees, used tools, and, in addition to being provisioned, ate wild foods including ants, small mammals and birds. In 1978, when water levels were low, they waded to the mainland. Most were recaptured and returned to the laboratory, but 1 or 2 females evaded capture and may have encountered wild conspecifics.
- Case 3: Bear Island, Georgia, USA (1972–73). Hand reared at the Yerkes Regional Primate Research Center (YRPC), 8 chimpanzees were moved to the island to establish a breeding colony for biomedical purposes. They were provisioned and provided with unheated shelters. The apes climbed trees and ate wild foods. As ambient temperatures occasionally fell below zero, this possibly caused 1 female to die of pneumonia. Of 3 females who conceived, 2 miscarried. The experiment was terminated in 1974, and the remaining apes moved to YRPC.
- Case 4: Niokolo Koba National Park, Senegal (1972–79). In conjunction with a programme set up in Gambia by British national Stella Brewer, 14 chimpanzees, mostly confiscated, plus some from English zoos were set free. The releases were the first outside an island, taking place in successive waves in the neighbouring country Senegal in Niokolo Koba National Park and around Mount Assirik. Most animals had been trained how to eat wild foods, build nests, use hammer stones to crack hard fruits and how to avoid predators and snakes. The 3 young chimpanzees, 3–7 yrs old, shipped over from England perished upon arrival. One died of heat exhaustion while being trucked to the site. The 2 others, once their transport crates were opened, immediately ran into the forest; they were found dead the next morning, likely from exhaustion and dehydration. Some rehabilitants were injured by wild residents, several disappeared. Consequently, survivors were moved to an island in the River Gambia (Case 5).

- Case 5: Baboon Islands, River Gambia National Park (1979–85). Supported by wildlife authorities, Brewer moved chimpanzees onto islands that constitute River Gambia National Park – first 8 survivors from Niokolo Koba (Case 4) and later another 16. The islets were devoid of chimpanzees, which were in any case extinct in the country. Commonly called 'Baboon Islands', there are baboons and nearly a complete spectrum of Gambia's mammalian fauna, such as hyenas, leopards, monkeys, antelopes and pigs. The arrangement was meant to be temporary until a secure mainland habitat could be found, which, however, never materialised. Thus, this community was later merged with another group (Case 6).

- Case 6: Baboon Island, River Gambia National Park (1979). In the same year that Brewer initiated the first island-release (Case 5), Janis Carter put 9 animals onto a neighbouring island, after training them in survival skills. Most rehabilitants had been confiscated. The impetus for the releases initially came from an additional ape, named Lucy, who was raised in the USA by Maurice and Jane Temerlin. Lucy was owned by the Institute for Primate Studies in Oklahoma and taught American Sign Language as part of an ape language project. Upon reaching adolescence, at age 12, she became disruptive, and it was thought her needs would best be served in the wild. Lucy was shipped to Gambia, accompanied by University of Oklahoma psychology graduate student Janis Carter. Initially, Lucy did not relate to other chimpanzees, instead preferring humans and showing signs of depression. Janis Carter lived with the apes on the island for several years, trying to acclimatise Lucy to the environment. The majority of the releasees survived, built nests and ate wild foods. One female captured, killed and ate a colobus monkey. The chimpanzees from this project were merged with Brewer's community in 1985 (Case 7).

- Case 7: Baboon Islands, River Gambia National Park (1985–2003). The island groups of Brewer (Case 5) and Carter (Case 6) were merged in 1985, under the name of Chimpanzee Rehabilitation Project. Animals were added until 2003 and are now found on 3 islands. Lucy (Case 6) was found dead in 1987. Attempts of oral contraception failed and many infants have been born. Caretakers cannot walk across the islets, as the chimpanzees might pose a danger. About three times per week, the apes are provisioned from boats. Following Brewer's death in 2008, the Gambian government took over the project, and the site is now solely managed by Carter. As of 2016, there were 116 animals. Some financial support is garnered from tourists who can view the apes from boats.

- Case 8: Little Bassa River, Liberia (1978–2006). In 1974, the New York Blood Center (NYBC) acquired a chimpanzee-keeping biomedical facility in Liberia and converted it into a research colony known as ViLab, aiming to develop vaccines. A large number of 200–400 apes were kept here, sourced as confiscated pets, orphans or taken from the wild. ViLab quickly became crowded, because many captives, often kept singly in cages, could only serve briefly as research subjects. Thus, driven by what the company described as 'welfare' concerns, apes were transferred onto 5–6 small islands near the ViLab facility. The releases of approximately 125 animals happened between 1978–2006, in 6 cohorts of about 18 (1978), 24 (1983), 22 (1985), 18 (1987), 30 (1987), and more than 15 (2005–06). Most, if not all, animals were at times relocated to ViLab itself and later released again, to treat illnesses, mitigate conspecific aggression, and, importantly, because Liberia, periodically from 1989 to about 2005, became embroiled in civil war. On the islets, numerous animals died from drowning, conspecific aggression or diseases, and during the periods of armed conflict, they perished from hunger, thirst, general neglect, even when returned to the lab. In 2007, ViLab was closed, and management of the islands transferred to the Liberian Institute for Biomedical Research. NYBC and the Humane Society of the United States are committed to the lifetime care of the ex-lab chimpanzees plus their offspring – given a lack of contraception (see also Case 9). By 2016, 66 apes remained on the islands.

- Case 9: Bandama River Island, Azagny National Park, Ivory Coast (1983). ViLab (Case 8) also brought 20 chimpanzees onto a river island in neighbouring Ivory Coast. The lab-subjects were infected with pathogens, a likely reason why wildlife authorities objected to setting them free in mainland forests. Within a year, gastrointestinal disease (shigella) killed 6 animals, while another 5 disappeared. Survivors were transferred to a neighbouring island. By the end of 1984, just 1 male and 1 female remained. They conceived 2 offspring, but by 2013, only the male (Ponso) was still alive. While the company had ended its support in 2003, local people, and one villager in particular, takes care of the remaining chimpanzees – since 2015 through funds raised by the non-profit organization 'Les Ami de Ponso'.
- Case 10: Kerfalya Island, Konkoure River, Guinea (1992). European expatriate Helen Dorkenoo of NGO Help Guinea transferred 12 of 24 confiscated and otherwise rescued chimpanzees onto an islet of the size of a football-pitch. While provisioned, within 2 years, 10 had died and or disappeared, and by 2000, the remaining 2 had also perished.
- Case 11: Kibale National Park, Uganda (1994). A confiscated former pet, 5-yr old female Bahati, was taken by staff of the Kibale Forest Chimpanzee Project to meet wild conspecifics. In the Kanyawara community of habituated chimpanzees, Bahati (whose name translates to 'luck' in KiSwahili) had friendly interactions with 28 individuals. However, after brief stints with the wild apes, Bahati, within 1 month, showed up 5 times in surrounding villages, begging for food. When she finally refused to go back to the forest, she was transferred to Entebbe zoo and then Ngamba (Case 13).

- Case 12: Isanga Island, Lake Edward, Uganda (1995–96). Likely related to crowding at Entebbe zoo, the Jane Goodall Institute transferred 12 chimpanzees onto the islet. Two days post-release, 2 male infants were found dead in the water. The apes were provisioned, often from a boat, and tourists could view them. By about 1998, all survivors were returned to the zoo, and the majority later moved to Ngamba (Case 13).
- Case 13: Ngamba Island, Lake Victoria, Uganda (1998). An initiative by Australian Debby Cox and a consortium of Ugandan wildlife authorities, national and international NGOs, 15 chimpanzees were moved to Ngamba. The animals are provisioned four times per day and enticed to join a holding facility each night. Ngamba, since 1999, is a popular tourist destination, with proceeds used to support the project. By 2022, the sanctuary held 52 animals, including recent additions following confiscations.
- Case 14: Afi Mountain Wildlife Sanctuary, Nigeria (2000). From the Pandrillus centre, which cares for about 30 rescued chimpanzees, about 4 escaped into the surroundings, where wild conspecifics live. While 1 was recaptured the next day, and another 1 after 3 months, 2 males survived in the wild for at least 3 months and were never recaptured; 1 was possibly killed by hunters.
- Case 15: Concouati-Douli National Park, Congo (1996–2001, 2005–12). French expatriate Aliette Jamart had founded a sanctuary in Pointe Noire in 1989 and in 1991 the non-profit HELP (Habitat Ecologique et Liberté des Primates). A year later, she moved chimpanzees to islets in Concouati-Douli National Park, where they had opportunity to acquire nest building and foraging skills. All future rehabilitants were health-checked and DNA-tested to ascertain their geographic origin. From 1996, over 5 years, 37 animals were let out into the forest, most of them radio-collared and extensively monitored. Wild conspecifics attacked almost all males and a third of all females. While veterinarians saved some victims, about 6 died. Several females joined wild groups. The remaining formed a chimpanzee-typical fission-fusion community. Many births occurred. Another 16 were released between 2005–12, presumably following the same protocol.
- Case 16: Tacugama Chimpanzee Sanctuary, Sierra Leone (1998, 2006). The sanctuary cares for confiscated and rescued chimpanzees. During walks in the surrounding forests of the Western Area Peninsula National Park, 5 very young individuals escaped. Of these, within 1 year, 3 were recaptured and 1 more seen in good health. Some of the escapees might have been fostered by a wild female. An additional 31 escaped in 2006, when a juvenile, using tools, opened the enclosure. A visitor to the sanctuary was severely injured, and the taxi driver who had brought the tourist was killed. Most chimpanzees – 21 – returned voluntarily, 6 were captured, 4 remain unaccounted for.

- Case 17: Diarra River, Senegal (2002). A local farmer came across a dead chimpanzee female, presumably the mother of an infant nearby. It took Janis Carter and the villager 1 month to obtain official permission to return the infant to its natal group. The infant was let out of a bag, seemingly accepted, and seen with other chimpanzees on several occasions thereafter.
- Case 18: Unknown location, Guinea (before 2013). Janis Carter freed 2 orphaned infants, cared for by humans for less than 6 months, near where they had been captured. Amongst excitement, the chosen chimpanzee group accepted the infants.
- Case 19: Haute Niger National Park, Guinea (2008, 2011). The Chimpanzee Conservation Centre (CCC) acclimatised 17 of its residents through forest walks and exposure to wild foods. After health-checks and genetic blood tests to assure they belonged to the local subspecies, they were fitted with radio-collars and set free. Several dispersed widely, were retrieved and released again, a process during which 1 anesthetized male died. Another male was slightly injured by wild conspecifics. A female releasee returned to the sanctuary in the company of a wild female. In 2013, 8 rehabilitants had formed a free-ranging group. As of 2017, several released chimpanzees have joined wild conspecifics.
- Case 20: Sanaga River islands, Douala-Edea Wildlife Reserve, Cameroon (2008, 2010, 2015). The non-profit association PAPAYE France maintains a sanctuary in the Douala-Edea Wildlife Reserve near the Sanaga River. Orphans were taken on walks to accustom them to a natural forest, until 16 of them, in three waves (2008, 2010, 2015), were moved onto two river islets. They are fed daily at the shore, from boats and by care-takers wading towards the beach. Sometimes, they kill and eat monkeys native to the islets. One male was killed by villagers, all other rehabilitants survived. In 2017, together with newborns, they numbered 20 individuals.
- Case 21: Fongoli Savanna Chimpanzee Project, Senegal (2009). A 9-months old female infant was confiscated from hunters who had captured her when hunting dogs injured her mother. The adult female escaped. Within a week of capture, the infant was brought back to her natal community, which is habituated to humans. Once let free, the infant climbed onto an approaching adolescent male. Soon reunited with her mother, the infant was seen nursing. The mother recovered and the infant survived.

Are there any patterns that emerge from these 21 releases that took place over 60 years, from 1966 to 2015, involving, as far as we can ascertain, 448 chimpanzees?

While the translocations occurred throughout the geographical range of *Pan troglodytes*, covering 13 nations, the vast majority (70.3%) was released in West Africa (Sierra Leone, Gambia, Senegal, Guinea, Liberia), followed by Central Africa (18.1%; Nigeria, Cameroon, Gabon, Congo) and East Africa (9.8%; Tanzania, Uganda), with an outlier (1.8%) in the USA. Given that West Africa constitutes the epicentre of export of chimpanzees, including large markets for ape bushmeat, this region might have deeply engrained practices of extracting apes from the wild and keeping them alive in human care – which could explain why there were so many candidates for release.

Information about the location at birth is available for 391 chimpanzees. Accordingly, 95.9% were wild-born and 4.1% captive-born. Coincidental or not, this is the exact same proportion of wild-born versus captive-born individuals as ascertained for the approximately 24,000 primates involved in the total of 234 translocations reviewed by Beck (2018).

Theoretically, wild-born releasees may possess skills they remember from infancy that could aid their survival once set free. Moreover, if the members of a translocation cohort stem from different natal groups, their varied skills portfolio could facilitate cross-individual learning and imitation (Ongman et al. 2013). However, to date, no such analyses have been conducted. Extrapolating the proportion of wild-born to the whole pool of 448 releasees translates into 428 wild-born chimpanzees. This tells us something about the overall scale of destruction inflicted upon wild populations. As we illustrated above, one wild-born animal in the care of humans is representative of at least 10 that lost their lives while this individual was captured. As a result, the collateral mortality associated with the capture of 428 individuals translates to a minimum of 4,300 dead apes.

The extreme overrepresentation of wild-born raises serious doubt about the persistent claim of many zoos that they are keeping and breeding apes for conservation-based reintroductions (e.g. Kölpin 2015) – as this simply has not and will not happen (Sommer 2021). In fact, probably no single captive-born chimpanzee ever switched to an autonomous life as a wild ape. The only translocation where this might have occurred is the one at Haut Niger National Park (Case 19), as it involved 2 captive-borns. However, while the fate of one of them is unknown, the other captive-born releasee returned voluntarily to the sanctuary. In any case, even if zoos would scale back their ambitions and re-dress their breeding programmes as working towards an 'ark' that preserves the last remaining animals of their type, it seems almost cynical to produce more captive apes when thousands of them linger in habitat-country sanctuaries – including almost 1,200 chimpanzees (cf. Table 6.2).

For 303 out of 448 animals, we know the sex. Accordingly, 44.9% were males and 55.1% females – the reverse of the proportion in sanctuaries, where there are 53% males and 47% females (see above). Given that males are more aggressive, this could explain why more of them end up in sanctuaries in the first place, as pet-owners might be keen to get rid of them. However, by the same token, caretakers should have an extra motivation to let go of males, to reduce trouble in the sanctuary – but this is not what we found. Conversely, sanctuary managers may honour great ape release criteria which stipulate that aggressively acting individuals should not be set free – and thus selectively retain their males. There is also some evidence that male orphans become particularly attached to human caretakers, which would render their successful translocation more difficult (Beck 2018: 44). Finally, the underrepresentation of translocated males might be a corollary of their greater sex-differential mortality throughout particularly the younger life-history stages (Havercamp et al. 2019). All things considered, we currently do not understand how an even sex-ratio at birth becomes biased towards males in sanctuaries and towards females in release operations.

For roughly 278 individuals, we have information or estimates about their age. Accordingly, 15.8% of the transferees were between 1 and 5 years, 40.6% between 6 and 10 years, 26.3% between 11 and 15 years and 10.1% between 16 and 20 years. Given that the vast majority were wild-born and thus captured as infants, almost all had spent lengthy periods 'behind bars'. In every third case, captivity lasted for more than 10 years. In addition, numerous apes, after short stints of 'freedom', were returned to their bleak lab-confinement (Cases 2, 3, 8).

Given that almost all releasees had been removed from the wild as infants, they likely lacked skills to construct night-nests, which might be beneficial in terms of thermoregulation, comfort or pathogen-control (Chapter 4). Indeed, early studies of wild (van Lawick-Goodall 1968; Baldwin et al. 1981) and orphaned chimpanzees (Bernstein 1969) suggested that nest building proficiency would come from observing conspecifics. Consequently, some caretakers have deemed it important to train future releasees. For example, Brewer (1978) provided elevated platforms 7–8 m above ground and cut branches as props for a leafy bedding, which arguably stimulated nest building habits (Case 4, 5). With her charges, Carter (1981), for several months, constructed day nests on or close to the ground (Case 6). However, the apes were later reluctant to switch to arboreal nests, preferring to sleep close to their human caregiver. Other translocations did not involve any instruction, and yet almost all animals, sooner or later, built sleeping platforms anyway (e.g. Cases 1, 2, 3, 8). Indeed, recent studies have found that captives, without prior training, will construct nests when provided with appropriate materials (Locke et al. 2011). Perhaps, there are only a few ways, in which one can bend branches so as to create a reasonably sturdy platform, and the road to success is paved with some simple trial-and-error learning. In any case, the Rubondo data indicate that, at least after a couple of generations, characteristics of nesting architecture as well as nest use fall within the variation observed for native chimpanzees (Chapter 4).

Similarly, liberated chimpanzees seem to be able to select – and obtain – wild foods without being prompted how to do so. This includes the at least occasional consumption of animal matter, such as insects (Cases 1, 2) or mammals (Cases 1, 2, 6, 20).

Another issue looming large is the xenophobic nature of chimpanzees which triggers aggression against conspecifics. In the wild, territorial conflicts and violent inter-group encounters are well documented (Goodall 1986; Wrangham & Peterson 1996; Wilson 2013). These clashes, with their often extremely brutal 'lethal raiding', are likely driven by resource competition (Silk 2014). Consequently, it is not surprising when the introduction of foreign individuals prompts aggression by the residents. The most dramatic chain of events was triggered when 37 animals were released in Congo (Case 15). At the time, the 20 km^2 release area was estimated to harbour only 5–7 other chimpanzees. However, 9 of the 10 released males and 9 out of 16 females were attacked – with 4 males and 2 females left dead. More would have perished without veterinary intervention.

Conspecific attacks are likely, but not inevitable – as illustrated by the release in Haut Niger National Park (Case 19), which led to only one minor injury of a male, and the case of an adolescent female brought into contact with wild chimpanzees in Uganda who, instead of being attacked, instantly became part of the social network (Case 11). This is surprising, not least because of what happened in Congo. The idea has been floated that, compared to males, females are more suitable for release because they tend to leave their natal community (Hannah & McGrew 1991). Thus, male chimpanzees might welcome immigrant females as potential breeding partners. However, this reasoning considers only one side of the equation, as resident females might view same-sex immigrants as resource competitors. In fact, there is no scarcity of reports from the wild about females attacking and also killing other females (Boesch et al. 2008, Wilson et al. 2014, Lowe et al. 2020).

Lastly, conflict with humans might endanger the life of released animals – and vice versa. On this account, apart from rangers shooting apes in self-defence on Rubondo (Case 1), locals killed freed apes in Gambia (Case 7) and Cameroon (Case 20). And, when the Tacugama residents broke out of their enclosure, a local taxi driver was maimed to death while the visitor he transported was severely injured (Case 16). Indeed, chimpanzees can become very aggressive and injurious towards people – whether prompted by perceived self-defence, by trying to wrestle something desirable from humans, or even by considering people, particularly human infants, as a source of meat for consumption (Goodall 1986, Newman & O'Connor 2009, Hockings et al. 2010, McLennan & Hockings 2016, Quammen 2019).

In the light of all this: Were the releases 'successful'? A higher-bar success requires "contribution to the establishment of a self-sustaining wild population" (Beck 2018: v). Clearly, returning young animals to natal groups (Cases 11, 17, 18, 21) fulfils this criterion, albeit the 5 youngsters involved represent only 0.9% of the releasees. No other translocation qualifies for the high-bar criterion, because even in Concouati-Douli (Case 15) and Haut Niger (Case 19), where animals survived in the wild and some integrated into native groups, they did not 'add' anything to what were already viable populations. And while the Rubondo introduction might have established a self-sustaining population, this is based on different subspecies set free outside their natural range – a caveat that we will discuss below. Thus, all 17 translocations that did not concern natal groups miss even the lower-bar goal of "supplementing a wild population that is under carrying capacity or not viable" (ibid.). On top of this, a considerable proportion (41.1%) of the projects were terminated or were changed over to a different site (Cases 2, 3, 4, 5, 6, 11, 12).

Per definition, a restocking of native populations, whether higher-bar or lower-bar, is already impossible if animals are set free on islands where they can't interact with wild comrades. Islands make up a whopping 57.1% of all release sites; if we exclude the 6 returns to natal groups plus escapes, the figure is 80%. Practical as islands are in terms of having the apes under some degree of benevolent control, they require long-term provisioning, because all – except Rubondo – were or are too small to offer sufficient foodstuffs. Sometimes, such as with the three River Gambia islets of which the largest measures 1.8 km^2, the need for perennial provisioning might have only become obvious after the fact (Case 7). In comparison, the translocation of close to 50 chimpanzees onto the tiny island of Ngamba – just 0.4 km^2 – was done with the full realisation that this would entail long-term food supplementation (Case 13). At the extreme, if the economical or logistical support structures of individual caretakers or release organisations falter – as happened so dramatically with ViLab (Cases 8, 9; Gorman 2015) – animals may simply starve to death. They would not even have the chance to try their luck outside the confined space allocated to them – as the Escobar-hippos did so impressively.

In sum, our review of the releases conveys a picture of idiosyncrasy. There was neither a coordinated system of how animals could or should be prepared for life in the wild or monitored post release, nor were standardised procedures in place for how to mitigate arising problems. Instead, what transpires is 'anything goes' – reflecting varied motivations, contexts and constraints of the caretakers.

Conservation or Welfare: A Simmering Divide

We have not yet considered one aspect of the 21 chimpanzee translocations: their intended purpose. Our tabulation (cf. Table 6.2) attributes 31 primary or secondary motivations, which translate into the following proportions: conservation 9.7% (n = 3), welfare 61.3% (n = 19), commercial 3.2% (n = 1), scientific 3.2% (n = 1), other 22.6% (n = 7).

One could question the paucity of 'scientific' purposes, as many events have led to publications. However, except for the well-planned introduction in Concouati-Douli, the documentation and subsequent dissemination of release-related findings were not part of the original pre-release planning. Anyhow, welfare considerations were by far the prevailing cause for a release: aiming to reduce the burden of caretaking and allowing captive residents to live better lives.

How does this tally compare against the *Best Practice Guidelines for the Re-introduction of Great Apes* (Beck et al. 2007)? Published as an 'Occasional Paper of the IUCN Species Survival Commission', the guidelines are concerned with conservation-related translocations, which – as per our own definition – can be broken down into reinforcement, reintroduction and introduction. The paper subsumes all three modes under the term 'reintroduction'; similar to Beck (2018), who uses the same term for all translocations.

Just the same, this "tool for conserving great apes and their natural habitats" (Beck et al. 2007: 7) explicitly aims to "establish self-sustaining populations of great apes in the wild" in an area once part of their "historic range", but from which they have been "extirpated or become extinct". This goal can be achieved by creating an altogether new population – "re-establishing an extinct wild population" – or by "supplementing a wild population that is under carrying capacity or not viable" (ibid., p. 10, 12). Importantly, as we will soon revisit, these guidelines are *not* concerned with efforts to offer individual captive apes better prospects of life.

The guidelines formulate a best-case scenario (ibid., p. 7f) for what is a multistep process to achieve the conservation objective of establishing self-sustaining populations of great apes.

The first set of steps is preparatory, requiring a proposal with "the project's background, objectives, methodology, schedule and budget [plus] quantifiable measures by which the project's success can be assessed". This proposal is to be scrutinised "by a multidisciplinary specialist advisory team" and external reviewers. The release site "should be within the historic range of the taxon" and "include sufficient suitable habitat to support a self-sustaining population". Any past threat, that drove the previous extinction, "should have been addressed and resolved". There has to be "secure long-term financial support for the project, and approval from all relevant governmental and regulatory agencies". Local governments and people have to "endorse" the project.

The second set of steps concerns the apes destined for release. They "should be assessed behaviourally, physically and genetically" so that they are "likely to survive". Those "with significant deficits in survival-critical knowledge and skills" are to be granted "sufficient rehabilitation and post-release support". After they have been "medically examined, quarantined, treated" and vaccinated, each ape "should be permanently identified".

The third set of steps concerns the actual release. Staff must be enrolled in "an occupational health programme", including training on zoonotic diseases and "hygienic husbandry practices". There must be funds that enable "post-release monitoring" for "at least one year". A "clearly understood plan" details interventions, for example, how to treat an injured ape. The "outcomes and cost-effectiveness" of the project need to be disseminated, and the results have to be subject to "periodic external evaluation".

By these measures, not a single one (!) of the 21 historical translocations was of 'best standard'. Nowhere was a previously extinct population
re-established, and neither was there any supplementing of a non-viable
population. Of course, most releases took place before the guidelines were
formulated. Still, it is highly doubtful that, even in the future, programmes
can be designed that adhere to all stipulations of the guidelines and suc-
ceed to 're-establish' new great ape populations that are self-sustaining. To
supplement "a wild population that is under carrying capacity or not viable" will always be tricky, if only because 'carrying capacity' can hardly be
assessed. Moreover, it is highly unlikely, that multiple anthropogenic threats
which caused a population to decline can be "addressed and resolved".
These guidelines, if anything, are therefore more an aspirational check-list
of things to consider. If their prescriptive character is taken at face-value,
future ape releases seem unrealistic.

However, Benjamin Beck points out another "major and disconcerting
disconnect", namely the fact that welfare considerations were behind 46%
of the hundreds of releases of primates of all types (Beck 2018: iv–v) –
while for chimpanzee releases, the preponderance is 61%. Yet, the great ape
guidelines treat such motivations as almost an afterthought, by acknowledg-
ing that "rescue/welfare releases" may have "a different primary objective".
Nevertheless, "they should adhere to these guidelines as closely as possible.
Projects that address the welfare of individual apes must also consider the
conservation of the species as a whole" (ibid., p. 12).

Beck retells how, at a 2006 'African Primate Reintroduction Workshop'
organised by PASA and other expert groups, he became aware of a "simmer-
ing divide" between "practitioners of primate welfare-based reintroductions
and primate conservationists" (ibid., p. iv). His presentation of the guidelines project was "implored" by many attendees – because, "how could
struggling sanctuaries afford to do guidelines-mandated comprehensive
pre-release habitat surveys, extensive pre-release training, and systematic
post-release monitoring with radiotelemetry?". Sanctuaries have to "feed
hundreds of hungry mouths every morning", they have to "provide shelter
and love to an endlessly growing number of physically and psychologically
damaged ape orphans", while trying "to raise funds to keep their operations
going" (ibid., p. v).

As a result, many translocations of primates in general or chimpanzees
specifically were and are driven by the necessity to somehow reduce the burden of keeping the animals in cages or small enclosures. One doesn't have to
be a prophet to predict that such welfare-considerations will always be the
primary motivation behind releases. Given such constraints, the organisers
may well argue that they adhered to the guidelines "as closely as possible".
On this note: While many programmes lack expertise or resources to conduct systematic post-release monitoring, managers may also, consciously
or unconsciously, not really want to know what happened. Because, if an
animal 'disappeared' post release, this could mean the possibility of a happy

integration into a wild group; while intensive surveying might establish that the releasee didn't make it.

Within this context, Beck raises the question whether welfare-driven releases indeed improve the mental and physical well-being of the involved animals. Do they really provide a 'second life'? He cites the example of a confiscated ex-pet, the female Bahati (Case 11), who, despite being accepted by a wild chimpanzee community, left the forest again and again to re-establish contact with humans. Ultimately, efforts to "re-unite her with her comrades" were terminated, and she was handed over to a zoo and then a sanctuary. Siding with what he calls the 'Bahati Principle', Beck (ibid., p. vi) writes: "In terms of health and wellbeing of many orphaned, rescued, or long-term captive primates, life in a supportive, well managed, captive environment is preferable to life in the wild". He points out that even well-executed releases are "rarely beneficial" for the individual primates. "Hunger, aggression, pain and suffering, or at least discomfort, and conflict with humans are common after reintroduction. Freedom comes with a price".

Other released chimpanzees, too, did not want to remain at the sites chosen for them and instead went back to their human caregivers – one of them even introducing a wild female to an existence controlled by humans (Cases 15, 19). Similarly, of 31 animals that escaped from a sanctuary, 21 returned on their own accord (Case 16). The assertion that 'freedom comes with a price' is not only evidenced by Bahati and these other returnees, but by the fate of Lucy. Captive-born in the United States, and raised as if she were a human child, she was trained in ASL (American Sign Language). Lucy was 12 when she was taken from her American human household to Gambia, where she expressed 'hurt' via sign language and seemed depressed (Carter 1981). Clearly, this human-raised chimpanzee preferred human apes to the company of conspecific apes. Similarly, the chimpanzees who were shot on Rubondo when they sought access to familiar human foods (Chapter 2) would probably have lived longer even under the cruel conditions they endured in captivity back in Europe.

Our colleague Alexandra Palmer has investigated cost–benefit approaches to orangutan R&R (rehabilitation and release) programmes through interviews with practitioners and scientists involved. Similar to chimpanzees, at least 1,250 orangutans are currently biting their time in sanctuaries. Palmer's book *Ethical Debates in Orangutan Conservation* reflects upon the 'the red ape dilemma', which a chapter title formulates as 'Kill, incarcerate, or liberate?' (Palmer 2020: 47–74). Considering financial constraints, the 'kill'-option – under the euphemism of 'euthanasia' (good death) – would be the most economical, while 'incarcerate' is costliest, given the expenses of continued care. In terms of animal rights, euthanasia is the worse option, incarcerate the second best, whereas 'liberate' sounds like the panacea. Unfortunately, 'liberation' is difficult to organise, because there is hardly any suitable forest left, where any releasee could live a newly granted free life.

How comforting it would be, if our mandate to make the best of a bad situation was unambiguous. Palmer identifies one group of voices that has an unwavering opinion: religious interviewees. These devout Muslims or Christians subscribe to the view that God has created creatures for various roles – and the destiny of animals is in the wild. Thus, to restore a primordial balance disturbed by human hubris, animals such as apes must play their part – and, whether they like it or not, return to the wilderness where they belong. Because "God put orangutans in the forest for a reason", and because "the forest needs them", they "should do their duty there" (Palmer 2020: 70).

Benjamin Beck (2018: 16) identifies the religious stances as promoting "wildness over welfare", while the 'Bahati Principle' promotes "welfare over wildness". He confesses, that, given a choice, he would stick to the Bahati principle. Similarly, Mary Ann Warren has argued that, while we should strive to offer apes a "qualified right to liberty", we should prioritise their rights to life and well-being over their right to freedom (Warren 2001, cit. in Palmer 2020: 73).

Still there are also secular voices who maintain that it can be justified to wager the lives of releasees. The reality is that, not counting accidental escapes and returns to natal groups, *all* chimpanzee translocations, without exception, entailed the death of releasees, through people, accidents, diseases or conspecific aggression. One could argue that these losses have to be weighed against the suffering in perpetual confinement and that a short life in the wild is better than a long and torturous existence behind bars. This is the concept of "death with dignity", which, if taken to its extreme, could lead to the "dumping" of orphans into the forest (Palmer 2020: 55). So, do caretakers act responsibly when they translocate their charges? In reality, given limited resources, they may have little choice but to gamble with the life of those they try so hard to save – accepting that not all 'unwitting travellers' will survive.

This inescapable dilemma has been brought about by various parties. There are the unscrupulous wheeler-dealers, who uproot highly intelligent and sentient creatures from their natural homes to acquire and exploit them: biomedical labs, behavioural researchers, pet owners, circuses, zoos. There are the hunters who make a killing, literally, from trafficking bushmeat. And there are people – like the authors of this book and its readers – who rely on computers and smartphones peppered with precious metals mined in what was prime forest, who consume palm oil via their food or toiletries and who live and work in buildings made sturdier with tropical timber, resources that can only be extracted by disenfranchising apes from their habitats. Sadly, even the most successful releases will not lead to structural changes in what are deeply troubled human–animal relations.

Clearly, if we fail to protect habitats and stopping the extraction of apes from the wild, chimpanzees will not have a future anyway. The conservationist Nick Salafsky thus put it aptly, when he likened the considerable efforts that go into maintaining sanctuaries and the sporadic releases as "rearranging deck chairs on the Titanic" (cit. in Davis 2005).

Hybrids: Burden or Bonus for Conservation?

Looming large in terms of judging the Rubondo releases and their success in establishing a breeding population are the origins of the founder animals from two subspecies – the West and Central African variety of chimpanzee (Chapter 2). How does the resulting hybridity of their progeny impact on prospects to maintain a self-reliant population? And how is the descendants' cross-bred status viewed by the conservation community? We approach these issues by delineating thoughts we recently elaborated upon under the header 'Hybrid apes in the Anthropocene' (Palmer et al. 2021).

IUCN release guidelines for great apes stipulate that subspecies should not be mixed because of the danger of *outbreeding depression* (Beck et al. 2007). This policy rests on the notion that introducing unrelated genetic material into a breeding line is disadvantageous, because it breaks up co-adapted gene complexes. Thereby, 'untested' combinations are created, with potentially harmful effects which may not manifest until the second generation. There is indeed empirical evidence that cross-breeding causes chromosomal incompatibilities and sterility in otherwise healthy hybrids (e.g. spider monkeys, Ralls et al. 2013; owl monkeys, Deboer 1982), or the loss of local adaptations (e.g. hybrid ibex females birthed their calves too early in the season, leading to mass deaths; Frankham 2010).

Conversely, a lack of admixture may reduce genetic diversity and thus cause *inbreeding depression* (Keller & Waller 2002, Frankham 2010). This may manifest itself in deleterious recessive traits, reduced heterozygosity and associated higher rates of disease and parasite load with consequently lower breeding success (e.g. in endangered birds such as the kakapo, White et al. 2015). Strategic hybridisation is one technique to mitigate inbreeding depression, as the crossing of individuals with no immediate common ancestors can confer *hybrid vigour* (Edmands 2007). This effect is exploited in industries of domestic pedigree dogs and livestock (e.g. 'black baldy' cattle; 'blue butt' pigs; 'Cornish-Rock' poultry). Similar techniques are practiced in conservation, e.g. with the inbred and isolated population of Florida panther (Whiteley et al. 2015).

Apart from such artificial selection, admixture of previously distinct lineages may actually be favoured by natural selection, allowing us to view hybridisation as a 'creative' force for evolution (Fredriksen 2016). For example, blended genes played a crucial role in the development of the vast array of cichlid fish species (Meier et al. 2018), events that had turned Lake Victoria into 'Darwin's Dreampond' (Chapter 3).

In transgressive segregation, a fresh hybrid lineage may colonize a novel niche (e.g. cross-breeding of the now-extinct Aurochs and Caucasian wisent produced the European bison; Węcek et al. 2017). Such *reticulate evolution* (Arnold 2008) is also evidenced by the naturally occurring admixture of primate subspecies, species and even genera (Arnold & Meyer 2006, Zinner et al. 2011), as in lemurs (*Eulemur*) and gibbons (*Hylobates*). In baboons (*Papio* spp.), some populations consist of >30%

hybrid individuals, and cross-breeding may happen with distinct genera (*Rungwecebus*, *Theropithecus*; Walker et al. 2019).

Consequently, hybridisation appears to be not only quite common, but often advantageous, while tangible substantiation of negative effects of outbreeding is sparce, at least across mammalian taxa (Edmands 2007). Still, concerns about outbreeding depression carry weight in the conservation literature (Hoffmann et al. 2015, Frankham 2015), leading to calls that crossbreeds should be prevented or culled.

The very idea that intermixing is harmful may have more to do with a historic legacy of concepts of purity and human-centred racist and eugenicist ideologies (Biermann & Mansfield 2014). Hybrids, real or imagined, break down conventional boundaries between taxonomic categories ('liger', 'tigon') and even between humans and animals (centaur, sphinx). Because such fusions challenge our ambiguity tolerance (Bryson et al. 2020), hybrids are often viewed as unsettling, even monstrous (Rutherford 2018). Related sentiments have likely also influenced systematics, the branch of science concerned with classification. Hence, traditional taxonomy tends to reflect 'essentialist' views, by which each single species is defined through an unambiguous property – its 'essence' (Zachos 2016). Such more or less unconscious way of thinking likely also underlies the influential biological species concept (BSC) (Mayr 1942) which treats species as real and discrete entities. In this view, which often invokes ideas of outbreeding depression, hybridisation is negative for individuals as well as their source populations. However, a key problem with species definitions generally is the attempt to impose a discrete ordering system onto continuous life forms and processes (Zachos 2016), a philosophically impossible task. Taxonomy will therefore always remain messy and debatable, notwithstanding its value for conversations about the diversity of life (Kirksey 2015).

Still, mainstream conservation philosophy will often highlight the exceptionality of a particular population of interest, arguing that its 'purity' reflects a unique evolutionary lineage which should be preserved for posterity (Biermann & Mansfield 2014). In addition, composite animals created by human intervention are deemed particularly undesirable (Paquet & Darimont 2010). Conservationists have therefore aimed to prevent intermixing between cultivated and wild forms (e.g. domestic and Scottish wildcats, Fredriksen 2016; domestic dogs and wolves, Peltola & Heikkilä 2018) or of native populations with introduced animals (e.g. released pet marmosets breeding with endemic *Callithrix* monkeys in Brazil, Malukiewicz 2019).

Furthermore, within a political context, scarcity is often valued over abundance in nature. In this way, highly endangered species garner greater attention in terms of conservation – creating a reason to divide rather than combine taxa (Mitchell 2016). For example, relatively few formerly unknown types of non-human primate were discovered during the last 50 years. Yet, rather loosely defined criteria relating to genetic markers, vocal communication, morphology, or behaviour led to already described primate taxa being split, resulting in an increase of primate species from 180 in

1967 to 480 by 2013 (Zinner & Roos 2016). As a case in point, in 1992, Madagascar was considered to harbour two species of the mouse lemur *Microcebus*; two decades later, the number of species had grown to at least 20 (Yoder et al. 2016), and may yet increase beyond the currently recognised 24 (Schüßler et al. 2020). All these new species can likely interbreed. Accordingly, humans sometimes create the very categorical prerequisites without which a 'hybridisation' would not occur in the first place.

Given such considerations, captive apes in particular became targets of policies concerning the hybrid status of individuals, albeit there is no consistent approach yet. For chimpanzee breeding programmes in the United States, subspecies identities are of no concern, either for the roughly 400 zoo-kept individuals (Steve Ross, studbook keeper for the Chimpanzee Species Survival Plan, pers. comm. to JNM) or in biomedical facilities (National Research Council Committee on Long-Term Care of Chimpanzees 1997). In contrast, since 2015, Japan's roughly 300 individuals have been separated into *P. t. verus* and all the rest; the two sections are not permitted to interbreed (Michael Huffman, pers. comm. to JNM). Similarly, the European Association of Zoos and Aquaria (EAZA) is making efforts to genetically identify the 728 chimpanzees in its accredited zoos. The 359 already tested individuals are separated into breeding pools for *P. t. verus* and *P. t. troglodytes*, while numbers of *P. t. schweinfurthii* and *P. t. ellioti* are considered too low to warrant coordinated action (Frands Carlsen and Tom de Jongh, coordinators for the chimpanzee EAZA Ex situ Programme, pers. comm. to JNM). Since the early 2000s, a breeding moratorium has already been imposed for 102 identified hybrids, and this is now extended to all newly identified hybrids or not-yet-tested animals. EAZA aims to reduce the number of hybrids, while growing the non-mixed populations. Still, a 2016 viability assessment established that, under current management plans, *P. t. verus* would decline in the short- to mid-term. Therefore, a few individuals that are not *P. t. verus* (with desirable behavioural skills or breeding experience) were selected to bolster the stock of the West African variety. For existing hybrids, the long-term plan is nevertheless to reduce them to zero.

Zoo breeding programmes are commonly criticised because captive populations are often not genetically viable and as there is a lack of suitable habitat to set captive-bred animals free (Braverman 2012). Certainly, the historical chimpanzee releases which we have reviewed above did little or nothing for species preservation, at least in relation to the comparably many more individuals surviving in the wild. There is therefore little reason to think that maintaining genetically 'pure' populations in captivity would be beneficial to individual apes or necessary for conservation. If zoos nevertheless aim to keep or restore 'pure' types, they value some individual apes more than others, just because of their genetic make-up – a stance highly problematic in terms of animal rights.

In contrast, as we move deeper into the Anthropocene, with humans increasingly shaping global climate and environments (Lewis & Maslin 2018), critics of anti-hybrid conservation policies propose that, to use Bruno

Latour's phrase, we should "love our monsters" (Latour 2011), i.e. care for the new creatures humans have generated (Rutherford 2018), instead of culling or sterilising them (Fredriksen 2016). Moreover, the increasingly pervasive effects of people on the non-human world and the concomitant decline of wildlife due to habitat destruction may anyway render it impossible to maintain enough genetic diversity within previously distinct taxa without passive (natural) or facilitated (anthropogenic) intermixing of types. Shifting environments and climates will necessarily mean that organisms will need to adapt or otherwise perish. Therefore, it might be time to consider genetic diversity as having a higher priority than attempts to preserve pure types. Indeed, following Australia's wildfires in 2020, assisted gene flow was suggested as a means to ensure the survival of koalas (Seddon & Schultz 2020). Accordingly, conservation in the Anthropocene may see more "post-natural" and future-oriented "experimental" approaches (Palmer et al. 2021). Clearly, Bernhard Grzimek, without realising it, was a bold pioneer of this new stance towards human–animal relations.

Do we have any indications whether and how hybridisation impacted upon the Rubondo apes? If so, then the effects are likely minor, because, at least in terms of phenotypic characteristics, subspecies differ little from each other (Chapter 4). Most noticeable is a variation in face colour, with often light brown in *P. t. schweinfurthii*, deep brown to black in *P. t. troglodytes*, both light and dark brown to black in *P. t. ellioti*, while the pink pigmentation of *P. t. verus* infants turns dark later on. Also, *P.t. schweinfurthii* have comparatively long hairs. All these variations might not be adaptations *per se* (Jenkins & Napier 1976). Similarly, the larger size of western chimpanzees reflects ontogenetic scaling rather than fundamental differences in shape (Shea 1984). However, eastern chimpanzees may retain unique genetic adaptations to viral pathogens (Schmidt et al. 2019).

The only discernible phenotypic feature of the Rubondo apes potentially related to hybridisation are their noticeably larger sizes and darker brown faces in adults compared to *P. t. schweinfurthii*, e.g. at Gombe National Park (JNM pers. obs.). Otherwise, 50+ years post release, with descendants of at least the first and second generation in place, there is no evidence for outbreeding depression. On the contrary, the population grew annually by around 3.3% since inception, comparatively high compared to native communities (Chapter 4).

To explore the issue of hybridity in the Rubondo apes further, we would need to collect a rather demanding data set – about e.g. infant mortality and female reproductive output combined with a genetic investigation into the degree of hybridisation in the adults. However, years if not decades would pass before we could assemble such information. Still, even in the improbable case of outbreeding depression, instead of being generally endangered and ultimately doomed, the number of Rubondo apes might simply increase at a slower pace.

This far, the releases on Rubondo were not subject to specific management measures, triggered by hybrid issues or otherwise. The 'Tanzania

Chimpanzee Conservation Action Plan 2018–2023' (Tanzania Wildlife Research Institute 2018) acknowledges that "in Tanzania, the historical chimpanzee range is known to decline [...] over the past few decades, with the current numbers estimated to be in the region of 2,500 individuals" (p. 8). The Rubondo apes are – rightly – not considered to be wild per se, as this "population is not native, but is derived from chimpanzees [...] that were introduced to the island over four decades ago". The document asserts that, because of "the island's low chimpanzee population density and high resource availability [...] this population is currently in a good status, although over the long-term, inbreeding and diseases remain concerns", to then make a rather vague recommendation: "In this context, guidelines for the management of chimpanzee sanctuaries (Schoene & Brend 2002) are likely the most appropriate way to ensure the ongoing welfare of these animals" (p. 42). What practical implications this might have is not clear, not least because there are no chimpanzee sanctuaries in Tanzania. However, the article cited highlights that education – a role often connected with sanctuaries – is important for in-situ conservation. At present, it is hard to imagine that the Tanzanian government, even if subscribing to the aforementioned and often ill-informed views about hybrids, might wish to exterminate the Rubondo chimpanzees or remove them from the island.

Without doubt, the Rubondo apes are not contributing to the (re)establishment of a wild population. Still, being one of a kind, they represent a prototype of a self-reliant post-natural population of Anthropocene apes worth conserving in their own right. Why should we not love these monsters?

The Rubondo Introductions: Between Muddle and Model

How does Rubondo – with the benefit of hindsight – compare against other primate releases and specifically those of chimpanzees elsewhere? Whatever was done right at the time and whatever was not: at least a few of the apes set free onto the island must have done well, because they produced descendants who embody well-adapted apes.

Let us first try to take stock of the reasons behind the operation. Over the years, Bernhard Grzimek, from his TV program to his publications and his correspondence, oscillated between a variety of purposes (Chapter 1, Chapter 2). These range from a species conservation motive (prospects of ape–human xenotransplantation poses a threat to wild populations); to welfare considerations (having lost their 'infant appeal' for the public, adult apes are a burden for zoos); to a tourism-driven rationale (visitors might flock to Rubondo to view its charismatic wildlife); to habitat protection (income from tourists will motivate people to preserve natural environments); to aesthetic reasons (the island will become a 'paradise' for wildlife). Some of these considerations overlap with reasons for why other mammals such as rhinoceroses, giraffes and elephants were moved to Rubondo – but chimpanzees were still special in that they were ex-captive, brought over from Europe, and did not occur in the immediate surrounding region.

In terms of conservation motives, which category do the Rubondo translocations fit best? Let us remember that *conservation* releases include *reinforcement* of existing native populations; *reintroduction*, where native populations have disappeared; or *introduction* of a threatened taxon outside its historic range. According to IUCN guidelines, islands not inhabited by a released taxon still count to be within its historic geographical distribution if one of the adjacent banks is within the known range (Beck et al. 2007). To meet the geographically closest wild conspecifics would require travel from Lake Victoria's western bank to Nyungwe National Park (Rwanda), 400 aerial km away. However, it would seem quite a stretch to invoke the 'adjacent bank' criterion in order to declare Rubondo part of the 'historic' range of chimpanzees. More importantly, the unit of conservation, following IUCN guidelines, is the subspecies – and the nearest chimpanzees belong to *P. t. schweinfurthii* and not the two western-central African subspecies the Rubondo apes represent. Thus, the categories of reinforcement and reintroduction must be discounted, which leaves 'introduction' as the proper term.

If we could travel back in time, and do it all over, what could have been done better?

The Rubondo experiment, at times criticised for its lack of pre-release rehabilitation efforts (Borner 1985, Huffman et al. 2008), clearly challenges the conviction that translocations need to be preceded by elaborate preparations – such as getting the future releasees accustomed to a life in the wild, like sleeping in the open and fending for themselves. No such opportunity was extended to the Rubondo arrivals. Their crates were unlatched and liberation meant to be chucked into the unknown. That was it. What helped these apes, or at least some of them, to survive, was sheer resilience, the remarkable ability of chimpanzees – perhaps not shared by other types of apes – to make-do in an entirely new environment. A human being with a similar biographical background of deprivation would have likely not lasted more than a few days in 'the wild', before succumbing to stress, depression, disease or hunger.

Be that as it may, unlike what happened on Rubondo, sanctuary workers often invest considerable resources to prepare their charges for a 'new life'. Still, at present, rehabilitation protocols are less evidence-based than intuitive (Seddon et al. 2007, Guy et al. 2014). The word 'rehabilitate' itself, or at least its practical implications, is not without problems – as there is some arrogance in the assumption that all that is required to turn a human-raised ape into a 'wild' ape is to 'teach' some rather undemanding skills, such as bending branches, chomping tree fruit and to be suspicious of snakes. While not involving apes, a recent study evaluated the translocation of several groups of howler monkeys. Accordingly, survival outcomes were unrelated to the duration an individual had spent in captivity as well as to the length of the rehabilitation process (Tricone 2018). For apes, no project has hitherto documented whether and how duration and techniques of rehabilitation influence survival. There is also a lack of understanding of the role of individual suitability, physiologically as well as psychologically, albeit it is hard to see how solid criteria could be developed, instead of leaving the decision-making to the gut feeling of sanctuary managers. To gather enough convincing data, we would have to compare two cohorts, one cluster of apes that were released without any prior rehabilitation to another group that had underwent rehabilitation. However, even then, sample sizes would be small and subject to stochastic variation – apart from the near impossibility to assemble cohorts of future releasees that embody corresponding upbringings and age–sex compositions before releasing them into comparably similar environments.

As a proxy, we could look at survival rates of non-rehabilitated apes. For this, a comparison of the Rubondo operations with other releases provides a simple lesson: rehabilitation isn't necessary. Because, whatever you do or don't do, you win some, and you lose some. Within the cohorts destined for Rubondo, at least 2–3 animals died – if 3, then almost one fifth (18.8%). There might have been other deaths we are not aware of (Chapter 2). The passing away of female Joséphine on board the steamer ferrying the apes from Europe to Africa was probably avoidable; a heat stroke in a compact shipping crate sounds like an accident that could have been averted. Still, three other and likewise young chimpanzees, shipped from Europe to West Africa, suffered a similar fate during release-operations in Senegal (Case 4).

Moreover, ex-captive apes will always pose a danger to humans, given their strength and aggressive nature. There is certainly a higher chance that captive apes known to have injured people will continue to do so once released – as was the case with Jimmy (No. 12, cf. Table 2.1). Yet, an ex-captive female without such reputation likewise sank her teeth into a human (Chapter 2). At the minimum, a fence should have been erected around lodgings and offices to prevent Rubondo chimpanzees from entering. In addition, if the releases had been done today the apes would have been fitted with tracking collars or, better, microchips. Such technology could pinpoint their whereabouts, including those of specimens known to be dangerous. The devices could also geo-fence the accommodations of rangers or researchers, triggering alerts when trouble-makers approach, and thus keep the apes at a safe distance. Of course, such measures would not have protected those who worked in the forest from potential attacks. Grzimek actually suggested employing "Kanonenschläge" (fireworks explosives) – a tactic used today to scare elephants.

This brings us to the lack of post-release monitoring back in the 1960s, which has likewise been criticised (Borner 1985, Huffman et al. 2008, Beck 2018). Apart from the fact that later releases elsewhere often also fell short on the front of record-keeping, sophisticated technologies we could rely on nowadays did not exist at the time of the Rubondo releases. Tracking the apes on foot would have been not only dangerous – even if trackers went on joint patrols – but also quite ineffective on an island the size of Rubondo. All too often, trackers would likely not have found the apes. In addition, post-release monitoring would have conflicted with the objective to 'rewild' the apes and get them to avoid humans, instead of searching them out. Perhaps, more importantly, the personnel brought in by the Frankfurt Zoological Society in the immediate post-release years were trained foresters (cf. Table 2.2), not animal behaviourists. They were not chiefly tasked to monitor the releasees – or any other of the seven types of introduced mammals – but to aid the Tanzanian wildlife authorities in building up infrastructure on Rubondo (Matschke 1974, 1976). Records of the fate of the released apes are therefore restricted to intermittent notes about where 'chimpanzees' had been seen, while individual identification was rare.

By 1974, when Gustl Anzenberger set out to study the chimpanzees (Müller & Anzenberger 1995), most founders plus those born on the island seemed to have largely lost their interest in interacting with humans; at least, Anzenberger found it difficult to find them, and switched to research on bees instead (Anzenberger 1977). Nowadays, all Rubondo chimpanzees avoid humans. Thus, ironically, recent attempts to habituate them for tourism and research have necessitated a reversal of the initial impetus to instil a shyness of humans in the apes.

More than anything, the relatively favourable outcome of the Rubondo releases is owed to the quality of the habitat in which they were set free. Conversely, other translocations faced an uphill battle from day one, because of limited food sources, conflict with resident conspecifics and humans. We could add 'unsuitable climate' to this list, given the experiment on Bear Island in Georgia, USA (Case 3). As the release site appears to largely determine the outcome, no place could have been more perfect than Rubondo island! While some destination sites are unbounded and others "confined by water" (Beck 2018: 2), such 'confinement' is certainly not felt on Rubondo. In fact, here, the apes can 'freely' move over a much larger area than primates at most forested mainland sites.

On that account, while the historic Rubondo releases clearly violate the later formulated IUCN guidelines from 2007, they are in remarkable compliance with the IUCN recommendations of 2013 about preferred sites for translocations of living plants and animals (IUCN/SSC 2013, cit. in Beck 2018: 3). In fact, in retrospect, the Rubondo releases meet *all* the rather stringent criteria of these guidelines, because this destination site was (i) large enough, (ii) contained sufficient resources, (iii) had an appropriate climate, (iv) was free from anthropomorphic threats and (v) had legal regulations permitting the introduction.

What would primatologists still want to know about the Rubondo apes? Some questions can only be answered through direct observations of individually identified animals, other avenues can be explored through the ever more sophisticated methods of molecular biology and genetics. Certainly, much progress could be made by habituating the chimpanzees to human observers. This would allow to ascertain whether the spectrum of material culture is indeed as sparse as we currently believe, and if there are customs or traditions that distinguish the Rubondo denizens as yet another unique 'ape ethnic group'. We would also be curious how the complete lack of natal dispersal, respectively immigration of animals from other communities, affects the social network. Is there less aggression from males upon females than typically observed at native sites? Are females more affiliative, because, after all, they are interrelated? Conversely, are there patterns of interaction that could be traced to intergenerational trauma inherited from the ex-captive founders? These observations could be supported by genetic analyses that determine paternity – as well as the population's deep history, by revealing the number of founders who contributed to the current population. Similarly, genetic research could compare allele frequency spectra of native chimpanzee subspecies to those of the Rubondo hybrids, helping us to understand which genetic traits have been particularly favoured in the Rubondo descendants.

Finally, what might the future hold for the Rubondo apes? The most important question is whether they will become a self-sustaining population. Its hybrid texture, owed to origins from different subspecies – and different locales – might well add to their overall resilience. There are no comparable communities of native apes that exist in the confines of an isolated space of just 240 km², we therefore simply do not know if the Rubondo dimensions are per se sufficient or not. Nevertheless, given that a community such as at Ngogo, Uganda, occupies just 28 km² of forest and numbers above 150 individuals (Watts et al. 2012), and that food sources on Rubondo are not currently constrained (Chapter 5), it will certainly take many decades, before Rubondo reaches its carrying capacity for the chimpanzees. Even then, that would not ring in their demise; instead, further growth would be limited by mammal-typical adjustments to reproduction, from lengthened inter-birth intervals to a later age of sexual maturity (MacArthur & Wilson 1967, Pianka 1970). However, latest at that point, the team of Rubondo-United will split into different communities, and inter-group hostilities are likely to ensue.

What potential dangers do the Rubondo apes face? Political instability is plaguing many African nations, and while Tanzania, since independence in the 1960s, belongs to the more stable category, the future is difficult to predict. Numerous apparently secure nations in Europe, Asia and Africa have descended into civil wars since the Rubondo releases took place. Should this fate befall Tanzania, no doubt, warring factions and militias will establish a foothold on the island, with loggers, miners and settlers following in tow – an all-too familiar unholy alliance that will destroy the island's wildlife, for trophies, for meat, for revenge, to demonstrate power (Benz & Benz-Schwarzburg 2010). More concrete are dangers from communicable diseases that can spill over from humans, from bats, from other mammals, from birds, from insects – the whole battery of yaws, anthrax or viruses such as Ebola, yellow fever, polio or SARS-CoV-2 (e.g. Leendertz et al. 2004, 2006, Calvignac-Spencer et al. 2012, Negrey et al.

2019). Such outbreaks could potentially wipe out the entire population, as happened to chimpanzees and gorillas during an Ebola-spell in Congo (Martínez et al. 2015).

Having said all this – at present, the Rubondo residents are one of a kind: apes that are, from their genetic make-up to the place they inhabit, synthetic creatures and thus a true product of the Anthropocene. And yet, they behave in such ways that a primatologist, not knowing anything about their history, would be bound to think they had been inhabiting that island for millennia. Nothing seems to be missing from the dazzling portfolio of behaviours that communities of chimpanzees are famous for and display at other sites.

Did Bernhard Grzimek achieve his ambitions on Rubondo? Scrutinised against the objectives Grzimek furnished for his 'Operation Chimpanzee', the answer has to be 'no'. Here is the tally: (i) species conservation – xenotransplantation did not become the envisioned threat; (ii) welfare considerations – relieving 17 chimpanzees from captivity while advocating their continued capture and exhibition behind bars is a bizarre contradiction; (iii) tourism – visitors do not flock to Rubondo to view its wildlife; (iv) habitat protection through revenue from tourism – the national park on Rubondo has never been profitable; (v) aesthetics – while some may view Rubondo as a 'paradise', others may see it as an example of distorted nature.

Without doubt, Grzimek's 'experiment' is embedded in an ethical minefield subsequent generations can clearly make out – notwithstanding tacit complicity. As a case in point, the authors of this book did not find it ethically reprehensible to seek out Rubondo as a study site. When working there, we enjoy what we see and are privileged to experience. With this, we are, intellectually and personally, beneficiaries of a late-colonial experiment and are therefore hardly on the moral high ground to pass judgement. Still, however entangled we are in the whole affair, we can identify lowlights and highlights and voice our opinion.

Is the world a better place because the Rubondo apes exist? Probably yes, if we forget about the humans that were disenfranchised to make room for the apes. Will this experiment be repeated? Probably not; but then, one day, an ape-loving multi-billionaire might tempt the islanders of neighbouring Maisome with a handsome financial compensation to resettle elsewhere; and replace humans with sanctuary apes.

At any rate, given habitat destruction on a global scale, it hardly matters that Grzimek tinkered with a couple of hundred square kilometers – where, as a result of his idiosyncratic interventions, large mammals still roam through what is now a remnant of rainforest. Be it as it may, in the moment, and hopefully for a long time to come, the artificial offshore sanctuary on Rubondo island embodies a chimpanzee haven.

Bibliography

ABEE, C.R., MANSFIELD, K., TARDIF, S. & MORRIS, T. (eds) (2012) *Nonhuman Primates in Biomedical Research* (2nd ed.). Academic Press, Boston MA

ADAMESCU, G.S., PLUMPTRE, A.J., ABERNETHY, K.A., POLANSKY, L., BUSH, E.R., CHAPMAN, C.A., et al. (2018) Annual cycles are the most common reproductive strategy in African tropical tree communities. *Biotropica* 50, 418–430

ALCOCK, J. (2009) *Animal Behavior: An Evolutionary Approach* (9th ed.). Sinauer Associates, Sunderland MA

ALMEIDA-WARREN, K., SOMMER, V., PIEL, A.K. & PASCUAL-GARRIDO, A. (2017) Raw material procurement for termite fishing tools by wild chimpanzees in the Issa valley, Western Tanzania. *American Journal of Physical Anthropology* 164, 292–304

AMMANN, K. & PEARCE, J. (1995) *Slaughter of the Apes: How the Tropical Timber Industry is Devouring Africa's Great Apes.* World Society for the Protection of Animals, London

AMMANN, K. & SOMMER, V. (1998) *Die Großen Menschenaffen: Orang-Utan, Gorilla, Schimpanse, Bonobo.* BLV, Munich

AMSLER, S.J. (2009) *Ranging Behaviour and Territoriality in Chimpanzees at Ngogo, Kibale National Park, Uganda.* PhD thesis. University of Michigan

ANDERSON, A.M. (1961) Further observations concerning the proposed introduction of Nile perch into Lake Victoria. *East African Agricultural and Forestry Journal* 26, 195–201

ANDERSON, D.P., NORDHEIM, E.V., BOESCH, C. & MOERMOND, T.C. (2002) Factors influencing fission-fusion grouping in chimpanzees in the Tai National Park, Côte d'Ivoire. In *Behavioural Diversity in Chimpanzees and Bonobos* (eds C. BOESCH, G., HOHMANN & L.F. MARCHANT), pp. 90–101. Cambridge University Press, Cambridge

ANDERSON, J.R., WILLIAMSON, E.A. & CARTER, J. (1983) Chimpanzees of Sapo Forest, Liberia: density, nests, tools and meat-eating. *Primates* 24, 594–601

ANGWAFO, T.E., ROLAND, A. & EBUA, V.B. (2018) Rehabilitation and reintroduction of wild born orphan chimpanzee (*Pan troglodytes*) within the Pongo and Okokong islands of the Douala-Edea wildlife reserve, littoral region Cameroon. *International Journal of Environment, Agriculture and Biotechnology* 3, 55–65

Anonymous (1966) Chimps with a problem. *The Standard* (Tanzania) (17 Jun)

ANZENBERGER, G. (1977) Ethological study of African carpenter bees of the genus *Xylocopa* (Hymenoptera, Anthophoridae). *Zeitschrift für Tierpsychologie* 44, 337–374

ANZENBERGER, G. (1986) How do carpenter bees recognize the entrance of their nests? An experimental investigation in a natural habitat. *Ethology* 71, 54–62

Arcus Foundation (2015) *Industrial Agriculture and Ape Conservation.* Cambridge University Press, Cambridge

Arcus Foundation (2018) *Infrastructure Development and Ape Conservation.* Cambridge University Press, Cambridge

ARNOLD, M.L. (2008) *Reticulate Evolution and Humans: Origins and Ecology.* Oxford University Press, Oxford

ARNOLD, M.L. & MEYER, A. (2006) Natural hybridization in primates: one evolutionary mechanism. *Zoology* 109, 261–276

AURELI, F., SCHAFFNER, C.M., BOESCH, C., BEARDER, S.K., CALL, J., CHAPMAN, C.A., et al. (2008) Fission-fusion dynamics: new research frameworks. *Current Anthropology* 49, 627–641

BALDWIN, P.J., SABATER PI, J., MCGREW, W.C. & TUTIN, C.E.G. (1981) Comparisons of nests made by different populations of chimpanzees (*Pan troglodytes*). *Primates* 22, 474–486

BALDWIN, P.J., MCGREW, W.C. & TUTIN, C.E.G. (1982) Wide-ranging chimpanzees at Mt. Assirik, Senegal. *International Journal of Primatology* 3, 367–385

BALTRUS, D.A. (2016) Divorcing strain classification from species names. *Trends in Microbiology* 24, 431–439

BASABOSE, A.K. (2004) Fruit availability and chimpanzee party size at Kahuzi montane forest, Democratic Republic of Congo. *Primates* 45, 211–219

BASABOSE, A.K. & YAMAGIWA, J. (2002) Factors affecting nesting site choice in chimpanzees at Tshibati, Kahuzi-Biega National Park: influence of sympatric gorillas. *International Journal of Primatology* 23, 263–282

BEENTJE, H.J. (1994) *Kenya Trees, Shrubs and Lianas.* National Museums of Kenya, Nairobi

BEENTJE, H.J. (2002) *Flora of Tropical East Africa – Pteridacea.* Royal Botanic Gardens Kew, CRC Press, London

BECK, B.B. (2010) Chimpanzee orphans: sanctuaries, reintroduction and cognition. In *Understanding Chimpanzees: The Mind of the Chimpanzee* (eds E.V. LONSDORF, S.R. Ross & T. MATSUZAWA), pp. 332–346. Chicago University Press, Chicago IL

BECK, B.B. (2018) *Unwitting Travelers: A History of Primate Reintroduction.* Salt Water Media, Berlin MD

BECK, B.B., WALKUP, K., RODRIGUES, M., UNWIN, S., TRAVIS, D. & STOINSKI, T. (2007) *Best Practice Guidelines for the Re-Introduction of Great Apes.* SSC Primate Specialist Group, Gland

BEHRINGER, V., DESCHNER, T., DEIMEL, C., STEVENS, J.M.G. & HOHMANN, G. (2014) Age-related changes in urinary testosterone levels suggest differences in puberty onset and divergent life history strategies in bonobos and chimpanzees. *Hormones and Behavior* 66, 525–533

BENZ, S. & BENZ-SCHWARZBURG, J. (2010) Great apes and new wars. *Civil Wars* 12, 395–430

BERNSTEIN, S. (1969) A comparison of nesting patterns among the three great apes. In *The Chimpanzee, Vol. 1: Anatomy, Behaviour and Diseases of Chimpanzees* (ed G.H. BOURNE), pp. 393–402. Karger, Basel

BERTOLANI, P. (2013) *Ranging and Travelling Patterns of Wild Chimpanzees at Kibale, Uganda: A GIS Approach.* PhD thesis. University of Cambridge

BEUNING, K.R.M., KELTS, K., RUSSELL, J. & WOLFE, B.B. (2002) Reassessment of Lake Victoria–Upper Nile River paleohydrology from oxygen isotope records of lake-sediment cellulose. *Geology* 30, 559–562

BIERMANN, C. & MANSFIELD, B. (2014) Biodiversity, purity, and death: conservation biology as biopolitics. *Environment and Planning D: Society and Space* 32, 257–273

BJORK, A., LIU, W., WERTHEIM, J.O., HAHN, B.H. & WOROBEY, M. (2011) Evolutionary history of chimpanzees inferred from complete mitochondrial genomes. *Molecular Biology and Evolution* 28, 615–623

BOESCH, C. (1996) Social grouping in Taï chimpanzees. In *Great Ape Societies* (eds W.C. MCGREW, L.F. MARCHANT & T. NISHIDA), pp. 101–113. Cambridge University Press, Cambridge

BOESCH, C. & BOESCH-ACHERMANN, H. (2000) *The Chimpanzees of the Taï Forest: Behavioural Ecology and Evolution.* Oxford University Press, Oxford

BOESCH, C., HOHMANN, G. & MARCHANT, L.F. (2002) *Behavioural Diversity in Chimpanzees and Bonobos.* Cambridge University Press, Cambridge

BOESCH, C., CROCKFORD, C., HERBINGER, I., WITTIG, R., MOEBIUS, Y. & NORMAND, E. (2008) Intergroup conflicts among chimpanzees in Taï National Park: lethal violence and the female perspective. *American Journal of Primatology* 70, 519–532

BOESCH, C. & WITTIG, R. (eds) (2019) *The Chimpanzees of the Taï Forest: 40 Years of Research.* Cambridge University Press, Cambridge

BOESCH-ACHERMANN, H. & BOESCH, C. (1994) The Tai chimpanzee project in Cote d' Ivoire, West Africa. *Pan Africa News* 1, 1–4

BORNER, M. (1980) Rubondo – ein Nationalpark mausert sich. *Das Tier* 10, 6–9

BORNER, M. (1985) The rehabilitated chimpanzees of Rubondo Island. *Oryx* 19, 151–154

BORNER, M. (1988) Translocation of 7 mammal species to Rubondo Island National Park in Tanzania. In *Translocation of Wild Animals* (eds L. NIELSEN & R.D. BROWN), pp. 117–122. Wisconsin Humane Society and Caesar Kleberg Wildlife Research Institute, Milwaukee WI

BOSCH, T.C.G. & MCFALL-NGAI, M.J. (2011) Metaorganisms as the new frontier. *Zoology* 114, 185–190

BRAVERMAN, I. (2012) *Zooland: The Institution of Captivity.* Stanford University Press, Stanford CA

BREWER, S. (1978a) *The Forest Dwellers.* Collins, London

BREWER, S. (1978b) *The Chimps of Mt. Asserik.* Alfred A. Knopf, New York NY

BREWER, S. (1980) Chimpanzee rehabilitation, why and how? *International Primate Protection League Newsletter* 7, 2–5

BREWER-MARSDEN, S., MARSDEN, D. & THOMPSON, M.E. (2006) Demographic and female life history parameters of free-ranging chimpanzees at the Chimpanzee Rehabilitation Project, River Gambia National Park. *International Journal of Primatology* 27, 391–410

BROCKMAN, D.K. & SCHAIK, C.P. VAN (eds) (2005) *Seasonality in Primates: Studies of Living and Extinct Human and Non-Human Primates.* Cambridge University Press, Cambridge

BRYSON, K., SOLIGO, C. & SOMMER, V. (2020) Interrogating boundaries against animals and machines: human speciesism in British newspapers. *Journal of Posthuman Studies* 4, 129–165

BUENO DE MESQUITA, C.P., NICHOLS, L.M., GEBERT, M.J., VANDERBURGH, C., BOCKSBERGER, G., LESTER, J.D., et al. (2021) Structure of chimpanzee gut microbiomes across tropical Africa. *mSystems* 6, e01269–20

CALDECOTT, J. & MILES, L. (2005) *World Atlas of Great Apes and Their Conservation.* University of California Press, Los Angeles CA

CALVIGNAC-SPENCER, S., LEENDERTZ, S.A.J., GILLESPIE, T.R. & LEENDERTZ, F.H. (2012) Wild great apes as sentinels and sources of infectious disease. *Clinical Microbiology and Infection* 18, 521–527

CAMPBELL, G., KUEHL, H., N'GORAN KOUAMÉ, P. & BOESCH, C. (2008) Alarming decline of West African chimpanzees in Côte d'Ivoire. *Current Biology* 18, R903–R904

CAMPBELL, T.P., SUN, X., PATEL, V.H., SANZ, C., MORGAN, D. & DANTAS, G. (2020) The microbiome and resistome of chimpanzees, gorillas, and humans across host lifestyle and geography. *The ISME Journal* 14, 1584–1599

CARLSEN, F. & DE JONGH, T. (2014) *European Studbook for the Chimpanzee Pan troglodytes.* Copenhagen Zoo, Frederiksberg

CARLSON, B.A., ROTHMAN, J.M. & MITANI, J.C. (2013) Diurnal variation in nutrients and chimpanzee foraging behavior. *American Journal of Primatology* 75, 342–349

CARTER, J. (1981) A journey to freedom. *Smithsonian* 12, 90–101

CARTER, J. (2003) Orphan chimpanzees in West Africa: Experiences and prospects for viability in chimpanzee rehabilitation. In *West African Chimpanzees: Status Survey and Conservation Action Plan* (eds R. KORMOS, C. BOESCH, M.I. BAKARR & T.M. BUTYNSKI), pp. 157–168. IUCN/SSC Primate Specialist Group, Gland/ Cambridge

CARVALHO, J.S., MEYER, C.F.J., VICENTE, L. & MARQUES, T.A. (2015) Where to nest? Ecological determinants of chimpanzee nest abundance and distribution at the habitat and tree species scale. *American Journal of Primatology* 77, 186–199

CHAPMAN, C.A., CHAPMAN, L.J., WANGHAM, R., HUNT, K., GEBO, D. & GARDNER, L. (1992) Estimators of fruit abundance of tropical trees. *Biotropica* 24, 527–531

CHAPMAN, C.A., CHAPMAN, L.J. & WRANGHAM, R.W. (1995) Ecological constraints on group size: an analysis of spider monkey and chimpanzee subgroups. *Behavioral Ecology and Sociobiology* 36, 59–70

CHAPMAN, C.A., CHAPMAN, L.J., WRANGHAM, R., ISABIRYE-BASUTA, G. & BEN-DAVID, K. (1997) Spatial and temporal variability in the structure of a tropical forest. *African Journal of Ecology* 35, 287–302

CHAPMAN, C.A., CHAPMAN, L.J., STRUHSAKER, T.T., ZANNE, A.E., CLARK, C.J. & POULSEN, J.R. (2005) A long-term evaluation of fruiting phenology: importance of climate change. *Journal of Tropical Ecology* 21, 31–45

CHAPMAN, C.A., CORRIVEAU, A., SCHOOF, V.A.M., TWINOMUGISHA, D. & VALENTA, K. (2017) Long-term simian research sites: significance for theory and conservation. *Journal of Mammalogy* 98, 652–660

CHOROWICZ, J. (2005) The East African rift system. *Journal of African Earth Sciences* 43, 379–410

CLAYTON, J.B., GOMEZ, A., AMATO, K., KNIGHTS, D., TRAVIS, D.A., BLEKHMAN, R., et al. (2018) The gut microbiome of nonhuman primates: lessons in ecology and evolution. *American Journal of Primatology* 80, e22867

CHONGHAILE, C. (2002) Abandoned chimps highlight conservation challenges in West Africa. *The Associated Press* (14 Sep)

Cox, D. (1998) Uganda: new homes and new projects. The Jane Goodall Institute World Report, 14–15

Cunningham, A.A. (1996) Disease risks of wildlife translocations. *Conservation Biology* 10, 349–353

Davis, K. (2005) Wild success, but at a cost. *New Scientist* (26 Feb), 18

Deboer, L.E.M. (1982) Karyological problems in breeding owl monkeys. *International Zoo Yearbook* 22, 119–124

Diehl, E. & Tuider, J. (eds) (2019) *Haben Tiere Rechte? Aspekte und Dimensionen der Mensch-Tier-Beziehung.* (bpb Schriftenreihe Vol. 10450). Bundeszentrale für politische Bildung, Bonn

Dinter, M. (2021) ZGF-Historie: Wie Grzimek die Schimpansen zurück nach Afrika brachte. *Gorilla (Magazin der Zoologischen Gesellschaft Frankfurt von 1858 e.V.)*, 3, 24–31

Dobzhansky, T. (1937) *Genetics and the Origin of Species.* Columbia University Press, New York NY

Dolezalova, J., Obornik, M., Hajduskova, E., Jirku, M., Petrzelkova, J.K., Bolechova, P., et al. (2015) How many species of whipworms do we share? Whipworms from man and other primates form two phylogenetic lineages. *Folia Parasitologica* 62, 63

Doran, D.M., McNeilage, A., Greer, D., Bocian, C., Mehlman, P. & Shah, N. (2002) Western lowland gorilla diet and resource availability: new evidence, cross-site comparisons, and reflections on indirect sampling methods. *American Journal of Primatology* 58, 91–116

Dore, K.M. (2017) Ethnophoresy. In *The International Encyclopedia of Primatology* (ed A. Fuentes), pp. 1–7. Wiley-Blackwell, Hoboken NJ

Dubiec, A., Gozdz, I. & Mazgajski, D. (2013) Green plant material in avian nests. *Avian Biology Research* 6, 133–146

Dutton, P.E., Chapman, H.M. & Moltchanova, E. (2014) Secondary removal of seeds dispersed by chimpanzees in a Nigerian montane forest. *African Journal of Ecology* 52, 438–447

Edmands, S. (2007) Between a rock and a hard place: evaluating the relative risks of inbreeding and outbreeding for conservation and management. *Molecular Ecology* 16, 463–475

Ely, J.J., Dye, B., Frels, W.I., Fritz, J., Gagneux, P., Khun, H.H., et al. (2005) Subspecies composition and founder contribution of the captive U.S. chimpanzee (*Pan troglodytes*) population. *American Journal of Primatology* 67, 223–241

Ewen, J.G., Armstrong, D.P., Parker, K.A. & Seddon, P.J. (2013) *Reintroduction Biology: Integrating Science and Management.* Wiley-Blackwell, Chichester

Fa, J.E., Currie, D. & Meeuwig, J. (2003) Bushmeat and food security in the Congo Basin: linkages between wildlife and people's future. *Environmental Conservation* 30, 71–78

Faith, J.T. (2014) Late Pleistocene and Holocene mammal extinctions on continental Africa. *Earth-Science Reviews* 128, 105–121

Farmer, K.H. (2002a) *The Behaviour and Adaptation of Chimpanzees (Pan troglodytes troglodytes) in the Republic of Congo.* PhD thesis. University of Stirling

Farmer, K.H. (2002b) Pan-African Sanctuary Alliance: status and range of activities for great ape conservation. *American Journal of Primatology* 58, 117–132

FAUST, L.J., CRESS, D., FARMER, K.H., ROSS, S.R. & BECK, B.B. (2011) Predicting capacity demand on sanctuaries for African chimpanzees (*Pan troglodytes*). *International Journal of Primatology* 32, 849–864

FERREIRA DA SILVA, M.J., KOPP, G.H., CASANOVA, C., GODINHO, R., MINHÓS, T., ZINNER, D. & BRUFORD, M.W. (2018) Disrupted dispersal and its genetic consequences: comparing protected and threatened baboon populations (*Papio papio*) in West Africa. *PLoS ONE* 13, e0194189

FISKEN, F.A., CARLSEN, F., ELDER, M., DE JONGH, T., PEREBOOM, J.J.M., POHL, B., et al. (2018) Global population records and managed-programme updates for the great apes: short report. *International Zoo Yearbook* 52, 212–226

FOLEY, C.A.H. & FAUST, L.J. (2010) Rapid population growth in an elephant *Loxodonta africana* population recovering from poaching in Tarangire National Park, Tanzania. *Oryx* 44, 205–212

FOWLER, A. (2006) *Behavioural Ecology of Chimpanzees (Pan troglodytes vellerosus) at Gashaka, Nigeria*. PhD thesis. University College London

FOWLER, A., KOUTSIONI, Y. & SOMMER, V. (2007) Leaf-swallowing in Nigerian chimpanzees: evidence for assumed self-medication. *Primates* 48, 73–76

FOWLER, A., PASCUAL-GARRIDO, A., BUBA, U., TRANQUILLI, S., AKOSIM, C., SCHÖNING, C. & SOMMER, V. (2011) Panthropology of the fourth chimpanzee: a contribution to cultural primatology. In *Primates of Gashaka: Socioecology and Conservation in Nigeria's Biodiversity Hotspot* (eds V. SOMMER & C. ROSS), pp. 503–544. Springer, New York NY

FRANDSEN, P., FONTSERE, C., NIELSEN, S.V., HANGHØJ, K., CASTEJON-FERNANDEZ, N., LIZANO, E., et al. (2020) Targeted conservation genetics of the endangered chimpanzee. *Heredity* 125, 15–27

FRANKHAM, R. (2010) Inbreeding in the wild really does matter. *Heredity* 104, 124

FRANKHAM, R. (2015) Genetic rescue of small inbred populations: meta-analysis reveals large and consistent benefits of gene flow. *Molecular Ecology* 24, 2610–2618

FREDRIKSEN, A. (2016) of wildcats and wild cats: troubling species-based conservation in the Anthropocene. *Environment and Planning D: Society and Space* 34, 689–705

FRUTH, B. (1995) *Nests and Nest Groups in Wild Bonobos (Pan paniscus): Ecological and Behavioral Correlates*. PhD thesis. Ludwig-Maximilians-Universität München

FRUTH, B. & HOHMANN, G. (1993) Ecological and behavioral aspects of nest building in wild bonobos (*Pan paniscus*). *Ethology* 94, 113–126

FRUTH, B. & HOHMANN, G. (1996) Nest building behavior in the great apes: the great leap forward? In *Great Ape Societies* (eds L.F. MARCHANT, T. NISHIDA & W.C. MCGREW), pp. 225–240. Cambridge University Press, Cambridge

FURUICHI, T. (2009) Factors underlying party size differences between chimpanzees and bonobos: a review and hypotheses for future study. *Primates* 50, 197–209

FURUICHI, T., HASHIMOTO, C. & TASHIRO, Y. (2001) Fruit availability and habitat use by chimpanzees in the Kalinzu Forest, Uganda: examination of fallback foods. *International Journal of Primatology* 22, 929–945

GAGNEUX, P., GONDER, M.K., GOLDBERG, T.L. & MORIN, P.A. (2001) Gene flow in wild chimpanzee populations: what genetic data tell us about chimpanzee movement over space and time. *Philosophical Transactions of the Royal Society of London. Series B, Biological Sciences* 356, 889–897

GETZ, W.M. & WILMERS, C.C. (2004) A local nearest-neighbor convex-hull construction of home ranges and utilization distributions. *Ecography* 27, 489–505

GHIGLIERI, M.P. (1984) *The Chimpanzees of Kibale Forest: A Field Study of Ecology and Social Structure*. Columbia University Press, New York NY

GILBY, I.C. & WAWRZYNIAK, D. (2018) Meat eating by wild chimpanzees (*Pan troglodytes schweinfurthii*): effects of prey age on carcass consumption sequence. *International Journal of Primatology* 39, 127–140

GISSIBL, B. (2019) *The Nature of German Imperialism*. Berghahn, New York NY

GOLDNER, C. (2014) *Lebenslänglich hinter Gittern. Die Wahrheit über Gorilla, Orang Utan und Co in deutschen Zoos*. Alibri, Aschaffenburg

GOLDSCHMIDT, T. (1996) *Darwin's Dreampond. Drama in Lake Victoria*. MIT Press, Cambridge MA

GONDER, M.K., LOCATELLI, S., GHOBRIAL, L., MITCHELL, M.W., KUJAWSKI, J.T., LANKESTER, F.J., et al. (2011) Evidence from Cameroon reveals differences in the genetic structure and histories of chimpanzee populations. *Proceedings of the National Academy of Sciences* 108, 4766–4779

GOODALL, J. (1986) *The Chimpanzees of Gombe: Patterns of Behaviour*. Harvard University Press, Cambridge MA

GORMAN, J. (2015) Chimpanzees in Liberia, used in New York Blood Center research, face uncertain future. *The New York Times* (28 May) [accessed 18 May 17]

GRZIMEK, B. (1962) *Rhinos Belong to Everybody*. Hill and Wang, New York NY

GRZIMEK, B. (1966a) Ein Platz für Tiere. TV episode, channel 1, ARD, West-Germany (47 min, 15 Nov)

GRZIMEK, B. (1966b) Apes travel from Europe to Africa. *African Wildlife* 20, 271–288

GRZIMEK, B. (1966c) Operation chimpanzee: youngster would not leave us. *Sunday News* (Tanzania) (18 Dec)

GRZIMEK, B. (1967) Operation chimpanzee: giant chimp starts dancing. *Sunday News* (Tanzania) (15 Jan)

GRZIMEK, B. (1969) Menschenaffen reisten von Europa nach Afrika. In *Grzimek unter Afrikas Tieren: Erlebnisse, Beobachtungen, Forschungsergebnisse* (ed. B. Brzimek), pp. 11–36. Ullstein, Berlin/Frankfurt a.M.

GRZIMEK, B. (1971) How Europe exported apes to Africa. In *Among Animals of Africa* (ed B. Grzimek), pp. 11–37. Harper Collins, London (USA edition 1970, Stein and Day, New York NY)

GRZIMEK, B. (1974) *Auf den Mensch gekommen: Erfahrungen mit Leuten*. Bertelsmann, Munich/Gütersloh/Vienna

GRZIMEK, B. (1988) Die Auswilderung von Schimpansen. In *Grzimeks Enzyklopädie Säugetiere Vol. 2* (ed B. GRZIMEK), pp. 482–485. Kindler, Munich

GRZIMEK, B. & GRZIMEK, M. (1959) *Serengeti darf nicht sterben. 367.000 Tiere suchen einen Staat*. Ullstein, Frankfurt a.M./Berlin/Wien

GRZIMEK, B., HAY, J.T. & HUTCHINS, M. (eds) (2003) *Grzimek's Animal Life Encyclopedia* [expanded from 13 to 17 volumes] (2nd ed.). Gale Cengage, Detroit MI

GUY, A.J., CURNOE, D. & BANKS, P.B. (2014) Welfare based primate rehabilitation as a potential conservation strategy: does it measure up? *Primates* 55, 139–147

HAKIZIMANA, D., HAMBUCKERS, A., BROTCORNE, F. & HUYNEN, M.C. (2015) Characterization of nest sites of chimpanzees (*Pan troglodytes schweinfurthii*) in Kibira National Park, Burundi. *African Primates* 10, 1–12

HAMILTON, W.D. (1971) Geometry for the selfish herd. *Journal of Theoretical Biology* 31, 295–311

HANNAH, A.C. & MCGREW, W.C. (1991) Rehabilitation of captive chimpanzees. In *Primate Responses to Environmental Change* (ed H.O. BOX), pp. 167–186. Chapman and Hall, London

HANSELL, M. (2007) *Built by Animals: The Natural History of Animal Architecture*. Oxford University Press, Oxford

HARRIS, D. & LUSTED, M.A. (2019) *The Transatlantic Slave Trade (Slavery in America)*. Essential Library, Minneapolis MI

HARRISON, M.E. & MARSHALL, A.J. (2011) Strategies for the use of fallback foods in apes. *International Journal of Primatology* 32, 531–556

HASEGAWA, H., IKEDA, Y., FUJISAKI, A., MOSCOVICE, L.R., PETRZELKOVA, K.J., KAUR, T. & HUFFMAN, M.A. (2005) Morphology of chimpanzee pinworms, Enterobius (*Enterobius*) anthropopitheci (Geldoelst, 1916) (Nematoda: Oxyuridae), collected from chimpanzees, *Pan troglodytes*, on Rubondo Island, Tanzania. *Journal of Parasitology* 91, 1314–1317

HAVERCAMP, K., WATANUKI, K., TOMONAGA, M., MATSUZAWA, T. & HIRATA, S. (2019) Longevity and mortality of captive chimpanzees in Japan from 1921 to 2018. *Primates* 60, 525–535

HEINSOHN, T. (2003) Animal translocation: long-term human influences on the vertebrate zoogeography of Australasia (natural dispersal versus ethnophoresy). *Australian Zoologist* 32, 351–376

HERBINGER, I., BOESCH, C. & ROTHE, H. (2001) Territory characteristics among three neighboring chimpanzee communities in the Tai National Park, Côte d'Ivoire. *International Journal of Primatology* 22, 143–167

HERNANDEZ-AGUILAR, R.A. (2006) *Ecology and Nesting Patterns of Chimpanzees (Pan troglodytes) in Issa, Ugalla, Western Tanzania*. PhD thesis. University of Southern California

HERNANDEZ-AGUILAR, R.A., MOORE, J. & PICKERING, T.R. (2007) Savanna chimpanzees use tools to harvest the underground storage organs of plants. *Proceedings of the National Academy of Sciences* 104, 19210–19213

HEYMANN, E.W. (2011) Florivory, nectarivory, and pollination: a review of primate-flower interactions. *Ecotropica* 17, 41–52

HICKS, A.L., LEE, K.J., COUTO-RODRIGUEZ, M., PATEL, J., SINHA, R., GUO, C., OLSON, S.H., SEIMON, A., SEIMON, T.A., ONDZIE, A.U., KARESH, W.B. et al. (2018) Gut microbiomes of wild great apes fluctuate seasonally in response to diet. *Nature Communications* 9, 1–18

HILL, K., BOESCH, C., GOODALL, J., PUSEY, A., WILLIAMS, J. & WRANGHAM, R. (2001) Mortality rates among wild chimpanzees. *Journal of Human Evolution* 40, 437–450

HIRAIWA-HASEGAWA, M. (1989) Sex differences in the behavioral development of chimpanzees at Mahale. In *Understanding Chimpanzees* (eds P.G. HELTNE & L.A. MARQUARDT), pp. 104–115. Harvard University Press, Cambridge MA

HLADIK, C.M. (1973) Alimentation et activité d'un groupe de chimpanzés réintroduit en forêt Gabonaise. *La Terre et Vie* 27, 343–413

HLADIK, C.M. (1977) Chimpanzees of Gabon and Chimpanzees of Gombe: some comparative data on the diet. In *Primate Ecology: Studies of Feeding and Ranging behaviour in Lemurs, Monkeys, and Apes* (ed T.H. CLUTTON-BROCK), pp. 81–501. Academic Press, London

HOCKINGS, K.J., YAMAKOSHI, G. & MATSUZAWA, T. (2017) Dispersal of a human-cultivated crop by wild chimpanzees (*Pan troglodytes verus*) in a forest–farm matrix. *International Journal of Primatology* 38, 172–193

HOCKINGS, K.J., YAMAKOSHI, G., KABASAWA, A., MATSUZAWA, T. (2010) Attacks on local persons by chimpanzees in Bossou, Republic of Guinea: long-term perspectives. *American Journal of Primatology* 72, 887–896

HOFFMANN, A., GRIFFIN, P., DILLON, S., CATULLO, R., RANE, R., BYRNE, M., et al. (2015) A framework for incorporating evolutionary genomics into biodiversity conservation and management. *Climate Change Responses* 2, 1

HOHMANN, G. & FRUTH, B. (2003) Culture in bonobos? Between-species and within-species variation in behavior. *Current Anthropology* 44, 563–571

HOHMANN, G., FOWLER, A., SOMMER, V. & ORTMANN, S. (2006) Frugivory and gregariousness of Salonga bonobos and Gashaka chimpanzees: the influence of abundance and nutritional quality of fruit. In *Feeding Ecology in Apes and Other Primates* (eds G. HOHMANN, M.M. ROBBINS & C. BOESCH), pp. 123–159. Cambridge University Press, Cambridge

HOMEWOOD, K., KRISTJANSON, P. & TRENCH, P.C. (2009) *Staying Maasai? Livelihoods, Conservation and Development in East African Rangelands.* Springer, New York NY

HUFFMAN, M.A., PAGE, J.E., SUKHDEO, M.V.K., GOTOH, S., KALUNDE, M.S., CHANDRASIRI, T. & TOWERS, G.H.N. (1996) Leaf-swallowing by chimpanzees: a behavioral adaptation for the control of strongyle nematode infections. *International Journal of Primatology* 17, 475–503

HUFFMAN, M.A. & CATON, J.M. (2001) Self-induced increase of gut motility and the control of parasitic infections in wild chimpanzees. *International Journal of Primatology* 22, 329–346

HUFFMAN, M.A., PETRŽELKOVÁ, K.L., MOSCOVICE, L.R., MAPUA, M.I., BOBÁKOVÁ, L., MAZOCH, V., et al. (2008) Introduction of chimpanzees onto Rubondo Island National Park Tanzania. *(SSC Re-Introduction Specialist Group, Abu Dhabi) Re-introduction NEWS (Special Issue)* 27, 213–216

HUFFMAN, M.A. & CHAPMAN, C.A. (eds) (2009) *Primate Parasite Ecology: The Dynamics of Host-Parasite Relationships.* Cambridge University Press, Cambridge

HUGHES, N., ROSEN, N., GRETSKY, N. & SOMMER, V. (2011) Will the Nigeria-Cameroon chimpanzee go extinct? Models derived from intake rates of ape sanctuaries. In *Primates of Gashaka. Socioecology and Conservation in Nigeria's Biodiversity Hotspot* (eds V. SOMMER & C. ROSS), pp. 493–523. Springer, New York NY

HUMLE, T., MAISELS, F., OATES, J.F., PLUMPTRE, A. & WILLIAMSON, E.A. (2016) *Pan troglodytes. The IUCN Red List of Threatened Species 2016*: e.T15933A102326672

HUNT, K. & MCGREW, W. (2002) Chimpanzees in the dry habitats at Assirik, Senegal, and at Semliki Wildlife Reserve, Uganda. In *Behavioural Diversity in Chimpanzees and Bonobos* (eds L. MARCHANT, C. BOESCH & G. HOHMANN), pp. 35–51. Cambridge University Press, Cambridge

HVILSOM, C., FRANDSEN, P., BØRSTING, C., CARLSEN, F., SALLÉ, B., SIMONSEN, B.T. & SIEGISMUND, H.R. (2013) Understanding geographic origins and history of admixture among chimpanzees in European zoos, with implications for future breeding programmes. *Heredity* 110, 586–593

ITOH, N. (2002) Food in the forests: chimpanzee food density and space distribution. In *The Mahale Chimpanzees: Thirty-Seven Years of Panthropology* (eds T. NISHIDA, S. UEHARA & K. KAWANAKA), pp. 77–100. Kyoto University Press, Kyoto

JANMAAT, K.R.L., BOESCH, C., BYRNE, R., CHAPMAN, C.A., GONÉ BI, Z.B., HEAD, J.S., et al. (2016) Spatio-temporal complexity of chimpanzee food: how cognitive adaptations can counteract the ephemeral nature of ripe fruit. *American Journal of Primatology* 78, 626–645

JANZEN, D.H. (1967) Synchronization of sexual reproduction of trees within the dry season in central America. *Evolution* 21, 620–637

JEAL, T. (2012) *Explorers of the Nile: The Triumph and Tragedy of a Great Victorian Adventure.* Faber and Faber, London

JENKINS, P.D. & NAPIER, P.H. (1976) *Catalogue of Primates in the British Museum (Natural History).* British Museum of Natural History, London

JENNRICH, R.I. & TURNER, F.B. (1969) Measurement of non-circular home range. *Journal of Theoretical Biology* 22, 227–237

JESUS, G.O.P. DE (2020) *Habitat Ecology and Primate Gregariousness in Nigeria's Gashaka Gumti National Park.* PhD thesis. University College London

JONES, C.G., LAWTON, J.H. & SHACHAK, M. (1994) Organisms as ecosystem engineers. *Oikos* 69, 373–386

KABASAWA, A., GARRIGA, R.M. & AMARASEKARAN, B. (2008) Human fatality by escaped *Pan troglodytes* in Sierra Leone. *International Journal of Primatology* 29, 1671–1685

KADE, U. (1967) Ich lebe mit den Schimpansen am Viktoriasee. *Das Tier* 6/7, 31

KAGORO-RUGUNDA, G. & HASHIMOTO, C. (2015) Fruit phenology of tree species and chimpanzees' choice of consumption in Kalinzu Forest Reserve, Uganda. *Journal of Ecology* 4, 477–490

KALBITZER, U., ROOS, C., KOPP, G.H., BUTYNSKI, T.M., KNAUF, S., ZINNER, D. & FISCHER, J. (2016) Insights into the genetic foundation of aggression in *Papio* and the evolution of two length-polymorphisms in the promoter regions of serotonin-related genes (5–HTTLPR and MAOALPR) in *Papionini. BMC Evolutionary Biology* 16, 121

KANO, T. (1983) An ecological study of the pygmy chimpanzees (*Pan paniscus*) of Yalosidi, Republic of Zaire. *International Journal of Primatology* 4, 1–31

KAPPELER, P.M. (1998) Nests, tree holes, and the evolution of primate life histories. *American Journal of Primatology* 46, 7–33

KAUR, T., SINGH, J., TONG, S., HUMPHREY, C., CLEVENGER, D., TAN, W., et al. (2008) Descriptive epidemiology of fatal respiratory outbreaks and detection of a human-related metapneumovirus in wild chimpanzees (*Pan troglodytes*) at Mahale Mountains National Park, Western Tanzania. *American Journal of Primatology* 70, 755–765

KELLER, L.F. & WALLER, D.M. (2002) Inbreeding effects in wild populations. *Trends in Ecology and Evolution* 17, 230–241

KELLERHOFF, S.F. (2015) Die "normale" Wahrheit hinter Grzimeks Nazi-Akte. *WELT-online* (04 Apr) [accessed 30 Dec 20]

KELLOGG, W.N. & KELLOGG, L.A. (1933) *The Ape and the Child: A Study of Environmental Influence Upon Early Behavior.* Whittlesey House, New York NY

KIERULFF, M.C.M., RUIZ-MIRANDA, C.R., DE OLIVEIRA, P.P., BECK, B.B., MARTINS, A., DIETZ, J.M., et al. (2012) The golden lion tamarin *Leontopithecus rosalia*: a conservation success story. *International Zoo Yearbook* 46, 36–45

KINGDON, J. (2021) *The Kingdon Pocket Guide to African Mammals* (2nd ed.). Princeton University Press, Princeton NJ

KINLOCH, B. (1972) *The Shamba Raiders: Memories of A Game Warden*. Collins/ Harvill Press, Hampshire/Ashford

KIRCHHOFF, C.A., WILSON, M.L., MJUNGU, D.C., RAPHAEL, J., KAMENYA, S. & COLLINS, D.A. (2018) Infanticide in chimpanzees: taphonomic case studies from Gombe. *American Journal of Physical Anthropology* 165, 108–122

KIRKSEY, E. (2015) Species: a praxiographic study. *Journal of the Royal Anthropological Institute* 21, 758–780

KIWANGO, Y.A., MWAMJENGWA, H.R. & NDAGA, S.R. (2005) *Rubondo Island National Park: Draft Report on Ethnoecology Study*. Tanzania National Parks. Unpublished Report

KÖLPIN, T. (2015) Sollten Tiere in Zoos gehalten werden? *Wilhelma Magazin* 3, 8–13

KÖNDGEN, S., KÜHL, H., N'GORAN, P.K., WALSH, P.D., SCHENK, S., ERNST, N., et al. (2008) Pandemic human viruses cause decline of endangered great apes. *Current Biology* 18, 260–264

KOOPS, K. (2011) *Elementary Technology of Foraging and Shelter in the Chimpanzees (Pan troglodytes verus) of the Nimba Mountains, Guinea*. PhD thesis. University of Cambridge

KOOPS, K., HUMLE, T., STERCK, E.H.M. & MATSUZAWA, T. (2007) Ground-nesting by the chimpanzees of the Nimba Mountains, Guinea: environmentally or socially determined? *American Journal of Primatology* 69, 407–419

KOOPS, K., MCGREW, W.C., MATSUZAWA, T. & KNAPP, L.A. (2012) Terrestrial nest-building by wild chimpanzees (*Pan troglodytes*): implications for the tree-to-ground sleep transition in early hominins. *American Journal of Physical Anthropology* 148, 351–361

KOUTSIONI, Y. & SOMMER, V. (2011) The bush as pharmacy and supermarket. Plant use by human and non-human primates at Gashaka. In *Primates of Gashaka. Socioecology and Conservation in Nigeria's Biodiversity Hotspot* (eds V. SOMMER & C. ROSS), pp. 135–230. Springer, New York NY

KREMER, W. (2014) Pablo Escobar's hippos: a growing problem. *BBC News Magazine* (26 Jun) [accessed 12 Jan 21]

KREYE, L. (2021) *"Deutscher Wald" in Afrika: Koloniale Konflikte um regenerative Ressourcen, Tansania 1892–1916. Umwelt und Gesellschaft, vol. 23* (eds C. MAUCH & H. TRISCHLER). Vandenhoeck & Ruprecht, Göttingen

KÜHL, H.S., KALAN, A.K., ARANDJELOVIC, M., AUBERT, F., D'AUVERGNE, L., GOEDMAKERS, A., et al. (2016) Chimpanzee accumulative stone throwing. *Scientific Reports* 6, 22219

KÜHL, H.S., BOESCH, C., KULIK, L., HAAS, F., ARANDJELOVIC, M., DIEGUEZ, P., et al. (2019) Human impact erodes chimpanzee behavioral diversity. *Science* 363, 1453–1455

LADYGINA-KOHTS, N.N. (2002) *Infant Chimpanzee and Human Child: A Classic 1935 Comparative Study of Ape Emotions and Intelligence*. (B. Wekker, Trans.) Oxford University Press, Oxford

LAMBERT, J.E. (1999) Seed handling in chimpanzees (*Pan troglodytes*) and redtail monkeys (*Cercopithecus ascanius*): implications for understanding hominoid and Cercopithecine fruit-processing strategies and seed dispersal. *American Journal of Physical Anthropology* 109, 365–386

LARGO, C.J., BASTIAN, M.L. & VAN SCHAIK, C.P. (2009) Mosquito avoidance drives selection of nest tree species in Bornean orang-utans. *Folia Primatologica* 80, 163

LATOUR, B. (2011) Love your monsters. In *Love Your Monsters: Postenvironmentalism and the Anthropocene* (eds M. SHELLENBERGER & T. NORDHAUS), pp 19–26. The Breakthrough Institute, Oakland CA

VAN LAWICK-GOODALL, J. (1968) The behaviour of free-living chimpanzees in the Gombe Stream Reserve. *Animal Behaviour Monographs* 1, 161–311

LACY, R.C. (1993) VORTEX: a computer simulation model for population viability analysis. *Wildlife Research* 20, 45–65

LEENDERTZ, F.H., ELLERBROK, H., BOESCH, C., COUACY-HYMANN, E., MÄTZ-RENSING, K., HAKENBECK, R., et al. (2004) Anthrax kills wild chimpanzees in a tropical rainforest. *Nature* 430, 451–452

LEENDERTZ, F.H., LANKESTER, F., GUISLAIN, P., NÉEL, C., DRORI, O., DUPAIN, J., et al. (2006) Anthrax in Western and Central African great apes. *American Journal of Primatology* 68, 928–933

LEHMANN, J. & BOESCH, C. (2003) Social influences on ranging patterns among chimpanzees (*Pan troglodytes verus*) in the Tai National Park, Côte d'Ivoire. *Behavioral Ecology* 14, 642–649

LEIGHTON, M. & LEIGHTON, D.R. (1982) The relationship of size of feeding aggregate to size of food patch: howler monkeys (*Alouatta palliata*) feeding in *Trichilia cipo* fruit trees on Barro Colorado Island. *Biotropica* 14, 81–90

LEKAN, T. (2016) A natural history of modernity: Bernhard Grzimek and the globalization of environmental *Kulturkritik*. *New German Critique* 128, 55–83

LEMOINE, S., PREIS, A., SAMUNI, L., BOESCH, C., CROCKFORD, C. & WITTIG, R.M. (2020) Between-group competition impacts reproductive success in wild chimpanzees. *Current Biology* 30, 312–318

LESNOFF, M. (2015) Uncertainty analysis of the productivity of cattle populations in tropical drylands. *Animal* 9, 1888–1896

LEWIS, S.L. & MASLIN, M.A. (2018) *The Human Planet How We Created the Anthropocene*. Yale University Press, New Haven CO

LITTLETON, J. (2005) Fifty years of chimpanzee demography at Taronga Park Zoo. *American Journal of Primatology* 67, 281–298

LOCKE, D.P., HILLIER, L.W., WARREN, W.C., WORLEY, K.C., NAZARETH, L.V, MUZNY, D.M., et al. (2011) Comparative and demographic analysis of orang-utan genomes. *Nature* 469, 529–533

LONSDORF, E.V., ROSS, S.R. & MATSUZAWA, T. (eds) (2010) *The Mind of the Chimpanzee*. University of Chicago Press, Chicago IL

LOWE, A.E., HOBAITER, C., ASIIMWE, C., ZUBERBÜHLER, K. & NEWTON-FISHER, N.E. (2020) Intra-community infanticide in wild, eastern chimpanzees: a 24-year review. *Primates* 61, 69–82

MACARTHUR, R.H. & WILSON, E.O. (2001 [1967]) The theory of island biogeography. *Acta Biotheoretica* 50, 133–136

MALUKIEWICZ, J. (2019) A review of experimental, natural, and anthropogenic hybridization in Callithrix marmosets. *International Journal of Primatology* 40, 72–98

MANN, A. & WEISS, M. (1996) Hominoid phylogeny and taxonomy: a consideration of the molecular and fossil evidence in an historical perspective. *Molecular Phylogenetics and Evolution* 5, 169–181

MARSHALL, B. (2018) Guilty as charged: Nile perch was the cause of the haplochromine decline in Lake Victoria. *Canadian Journal of Fisheries and Aquatic Sciences* 75, 1542–1559

MARTÍNEZ, M.J., SALIM, A.M., HURTADO, J.C. & KILGORE, P.E. (2015) Ebola virus infection: overview and update on prevention and treatment. *Infectious Diseases and Therapy* 4, 365–390

MATSCHKE, W. (1974) *Annual report 1973 of the Rubondo Island Game Reserve.* Game division Mwanza, Rubondo Island Game Reserve, Geita

MATSCHKE, W. (1976) Rubondo – Erinnerungen an eine Naturschutzinsel. *Der Deutsche Forstmann* 2, 24–25

MATSUMOTO-ODA, A. (2000) Chimpanzees in the Rubondo Island National Park, Tanzania. *Pan Africa News* 7, 16–17

MATSUMOTO-ODA, A., HOSAKA, K., HUFFMAN, M.A. & KAWANAKA, K. (1998) Factors affecting party size in chimpanzees of the Mahale Mountains. *International Journal of Primatology* 19, 999–1011

MATSUZAWA, T. (1991) Nesting cups and meta-tool in chimpanzees. *Behavioral and Brain Sciences* 14, 570–571

MATSUZAWA, T., HUMLE, T. & SUGIYAMA, Y. (2011) *The Chimpanzees of Bossou and Nimba.* Springer, New York NY

MAYR, E. (1942) *Systematics and the Origin of Species.* Columbia University Press, New York NY

MBARIA, J. & OGADA, M. (2016) *The Big Conservation Lie: The Untold Story of Wildlife Conservation in Kenya.* Lens and Pens, Auburn WA

MCCARTHY, M.S., LESTER, J.D. & STANFORD, C.B. (2017) Chimpanzees (*Pan troglodytes*) flexibly use introduced species for nesting and bark feeding in a human-dominated habitat. *International Journal of Primatology* 38, 321–337

MCCARTHY, M.S., DESPRÉS-EINSPENNER, M.-L., SAMUNI, L., MUNDRY, R., LEMOINE, S., PREIS, A., et al. (2018) An assessment of the efficacy of camera traps for studying demographic composition and variation in chimpanzees (*Pan troglodytes*). *American Journal of Primatology* 80, e22904

MCCULLOCH, B. & ACHARD, P.L. (1965) Rhino reserve in Lake Victoria. *Oryx* 8, 162–163

MCCULLOCH, B. & ACHARD, P.L. (1969) Mortalities associated with the capture, translocation, trade and exhibition of black rhinoceroses *Diceros bicornis.* *International Zoo Yearbook* 9, 184–191

MCGREW, W.C. (1992) *Chimpanzee Material Culture: Implications for Human Evolution.* PhD thesis. Cambridge University

MCGREW, W. (2004) *The Cultured Chimpanzee: Reflections on Cultural Primatology.* Cambridge University Press, Cambridge

MCGREW, W.C., MARCHANT, L.F., BEUERLEIN, M.M., VRANCKEN, D., FRUTH, B. & HOHMANN, G. (2007) Prospects for bonobo insectivory: Lui Kotal, Democratic Republic of Congo. *International Journal of Primatology* 28, 1237–1252

MCLENNAN, M.R. & HOCKINGS, K.J. (2016) The aggressive apes? Causes and contexts of great ape attacks on local persons. In *Problematic Wildlife: A Cross-Disciplinary Approach* (ed F.M. ANGELICI), pp. 373–394. Springer, Cham

MEHLHORN, H. (1988) *Parasitology in Focus.* Springer, Berlin/Heidelberg

MEIER, J.I., MARQUES, D.A., WAGNER, C.E., EXCOFFIER, L. & SEEHAUSEN, O. (2018) Genomics of parallel ecological speciation in Lake Victoria cichlids. *Molecular Biology and Evolution* 35, 1489–1506

MESHAKA, W.E.J., BUTTERFIELD, B.P. & HAUGE, J.B. (2004) *The Exotic Amphibians and Reptiles of Florida.* Krieger, Malabar FL

MIERSCH, M. (2007) Bernhard Grzimek – Deutschlands erster Umwelt-Held. *Die Welt* (07 Mar) [accessed 30 Dec 20]

MITANI, J.C., WATTS, D.P. & LWANGA, J.S. (2002) Ecological and social correlates of chimpanzee party size and composition. In *Behavioural Diversity in Chimpanzees and Bonobos* (eds C. BOESCH, G. HOHMANN & L. MARCHANT), pp. 102–111. Cambridge University Press, Cambridge

MITANI, J.C., WATTS, D.P. & AMSLER, S.J. (2010) Lethal intergroup aggression leads to territorial expansion in wild chimpanzees. *Current Biology* 20, 507–508

MITCHELL, A. (2016) Beyond biodiversity and species: problematizing extinction. *Theory, Culture and Society* 33, 23–42

MNAYA, B. & WOLANSKI, E. (2002) Water circulation and fish larvae recruitment in papyrus wetlands, Rubondo Island, Lake Victoria. *Wetlands Ecology and Management* 10, 131–141

MOELLER, A.H., PEETERS, M., NDJANGO, J.B., LI, Y., HAHN, B.H. & OCHMAN, H. (2013) Sympatric chimpanzees and gorillas harbor convergent gut microbial communities. *Genome Research* 23, 1715–1720

MOELLER, A.H., FOERSTER, S., WILSON, M.L., PUSEY, A.E., HAHN, B.H. & OCHMAN, H. (2016) Social behavior shapes the chimpanzee pan-microbiome. *Science Advances* 2, e1500997

MOSCOVICE, L.R. (2006) *Behavioral Ecology of Chimpanzees (Pan troglodytes) on Rubondo Island Tanzania: Habitat, Diet, Grouping and Ranging at a Release Site.* PhD thesis. University of Wisconsin-Madison

MOSCOVICE, L.R., ISSA, M.H., PETRZELKOVA, K.J., KEULER, N.S., SNOWDON, C.T. & HUFFMAN, M.A. (2007) Fruit availability, chimpanzee diet, and grouping patterns on Rubondo Island, Tanzania. *American Journal of Primatology* 69, 487–502

MOSCOVICE, L.R., MBAGO, F., SNOWDON, C.T. & HUFFMAN, M.A. (2010) Ecological features and ranging patterns at a chimpanzee release site on Rubondo Island, Tanzania. *Biological Conservation* 143, 2711–2721

MSINDAI, J.N. (2018) *Chimpanzees of Rubondo Island: Ecology and Sociality of a Reintroduced Population.* PhD thesis. University College London

MSINDAI, J.N., SOMMER, V. & ROOS, C. (2015) The chimpanzees of Rubondo Island: genetic data reveal their origin. *Folia Primatologica* 86, 327

MSINDAI, J.N., ROOS, C., SCHÜRMANN, F. & SOMMER, V. (2021) Population history of chimpanzees introduced to Lake Victoria's Rubondo Island. *Primates* 62, 253–265

MSINDAI, J.N. (2021) Ein Experiment mit gutem Ausgang: Die Schimpansen erobern Rubondo. *Gorilla (Magazin der Zoologischen Gesellschaft Frankfurt von 1858 e.V.),* 3, 32–36

MUELLER, N.T., BAKACS, E., COMBELLICK, J., GRIGORYAN, Z. & DOMINGUEZ-BELLO, M.G. (2015) The infant microbiome development: mom matters. *Special Issue: Nurturing the Next Generation, Trends in Molecular Medicine* 21, 109–117

MÜLLER, G. (1995) *Schimpansen auf Rubondo-Island, Tanzania: eine Pilotstudie.* Anthropologisches Institut, Universität Zürich. Unpublished Report

MÜLLER, G. & ANZENBERGER, G. (1995) *Chimpanzees (Pan troglodytes) of Rubondo Island, Tanzania. Anthropologisches Institut, Universität Zürich.* (Pilot study, German with English summary) https://doi.org/10.13140/RG.2.2.31713.12648 [www.researchgate.net; posted online in 2020]

MULLER, M.N. & WRANGHAM, R.W. (2014) Mortality rates among Kanyawara chimpanzees. *Journal of Human Evolution* 66, 107–114

MWAMBOLA, S., IJUMBA, J., KIBASA, W., MASENGA, E., EBLATE, E. & MUNISHI, L. (2016) Population size estimates and distribution of the African elephant using the dung surveys method in Rubondo Island National Park, Tanzania. *International Journal of Biodiversity and Conservation* 8, 113–119

MWANUZI, F.L. (2006) *Tanzania National Water Quality Synthesis Report*. Lake Victoria Environmental Management Project (LVEMP)

NAKAMURA, M., HOSAKA, K., ITOH, N. & ZAMMA, K. (eds) (2015) *Mahale Chimpanzees: 50 Years of Research*. Cambridge University Press, Cambridge/ New York NY

NAPIER, J.R. & NAPIER, P.H. (1967) *A Handbook of Living Primates*. Academic Press, New York NY

National Research Council Committee on Long-Term Care of Chimpanzees (1997) *Chimpanzees in Research: Strategies for Their Ethical Care, Management, and Use*. National Academies Press, Washington DC

NEGREY, J.D., REDDY, R.B., SCULLY, E.J., PHILLIPS-GARCIA, S., OWENS, L.A., LANGERGRABER, K.E., et al. (2019) Simultaneous outbreaks of respiratory disease in wild chimpanzees caused by distinct viruses of human origin. *Emerging Microbes and Infections* 8, 139–149

NEKARIS, K.A.I. & BERGIN, D. (2017) Primate trade (Asia). In *The International Encyclopedia of Primatology* (ed A. FUENTES), pp. 1–8. Wiley-Blackwell, Hoboken, NJ

NELSON, F. (2004) *The Evolution and Impacts of Community-Based Ecotourism in Northern Tanzania*. International Institute for Environment and Development, London

NEWMAN, A. & O'CONNOR, A. (2009) Woman mauled by chimp is still in critical condition. *The New York Times* (17 Feb) [accessed 12 Jun 21]

NEWTON-FISHER, N.E. (2003) The home range of the Sonso community of chimpanzees from the Budongo Forest, Uganda. *African Journal of Ecology* 41, 150–156

NEWTON-FISHER, N.E., REYNOLDS, V. & PLUMPTRE, A.J. (2000) Food supply and chimpanzee (*Pan troglodytes schweinfurthii*) party size in the Budongo Forest Reserve, Uganda. *International Journal of Primatology* 21, 613–628

NEWTON-FISHER, N.E. (2015) The hunting behavior and carnivory of wild chimpanzees. In *Handbook of Paleoanthropology* (eds W. HENKE & I. TATTERSALL), pp. 1661–1691. Springer, Berlin/Heidelberg

NISHIDA, T. (1968) The social group of wild chimpanzees in the Mahale mountains. *Primates* 19, 167–224

NISHIDA, T. (2011) *Chimpanzees of the Lakeshore: Natural History and Culture at Mahale*. Cambridge University Press, Cambridge

NISHIE, H. & NAKAMURA, M. (2018) A newborn infant chimpanzee snatched and cannibalized immediately after birth: implications for "maternity leave" in wild chimpanzee. *American Journal of Physical Anthropology* 165, 194–199

NISSEN, H.W. (1931) A field study of the chimpanzee. Observations of chimpanzee behavior and environment in Western French Guinea. *Comparative Psychology Monographs* 8, 1–122

NYANGANJI, G., FOWLER, A., MCNAMARA, A. & SOMMER, V. (2010) Monkeys and apes as animals and humans: ethno-primatology in Nigeria's Taraba Region. In

Primates of Gashaka: Socioecology and Conservation in Nigeria's Biodiversity Hotspot (eds V. Sommer & C. Ross), pp. 101–134. Springer, New York NY

Oates, J.F., Groves, C.P. & Jenkins, P.D. (2009) The type locality of *Pan troglodytes vellerosus* (Gray, 1862), and implications for the nomenclature of West African chimpanzees. *Primates* 50, 78–80

Obiero, K., Raburu, P., Okeyo-Owuor, J. & Raburu, E. (2012) Community perceptions on the impact of the recession of Lake Victoria waters on Nyando Wetlands. *Scientific Research and Essays* 7, 1647–1661

Odugbemi, T. (2008) *A Textbook of Medicinal Plants in Nigeria*. University of Lagos Press, Lagos

Ogutu-Ohwayo, R. (1990) The decline of the native fishes of lakes Victoria and Kyoga (East Africa) and the impact of introduced species, especially the Nile perch, *Lates niloticus*, and the Nile tilapia, *Oreochromis niloticus*. *Environmental Biology of Fishes* 27, 81–96

Ommaney, D. (1998) *Chimpanzee Status Survey and Habituation Project: Progress Report 1998*. Frankfurt Zoological Society. Unpublished Report

Ongman, L., Colin, C., Raballand, E. & Humle, T. (2013) The "super chimpanzee": the ecological dimensions of rehabilitation of orphan chimpanzees in Guinea, West Africa. *Animals* 3, 109–126

Opler, P., Baker, H.G. & Frankie, G. (1992) Seasonality of climbers: a review and example from Costa Rican dry forest. In *The Biology of Vines* (eds F.E. Putz & H.A. Mooney) pp. 377–392. Cambridge University Press, Cambridge

Osborne, C.P., Chuine, I., Viner, D. & Woodward, F.I. (2000) Olive phenology as a sensitive indicator of future climatic warming in the Mediterranean. *Plant, Cell and Environment* 23, 701–710

Pakenham, T. (1992) *The Scramble for Africa, 1876–1912*. Abacus, London

Palmer, A. (2020) *Ethical Debates in Orangutan Conservation*. Routledge, Abingdon

Palmer, A., Sommer, V. & Msindai, J.N. (2021) Hybrid apes in the anthropocene: burden or asset for conservation? *People and Nature* 3, 573–586

Pan African Sanctuary Alliance (PASA) (2019) *2019 Census of African Primates*. [pasa.org/wp-content/uploads/2019/12/2019_PASA_Census_Report_Final.pdf; accessed 12 Feb 21]

Paquet, P. & Darimont, C. (2010) Wildlife conservation and animal welfare: two sides of the same coin? *Animal Welfare* 19, 177–190

Paracer, S. & Ahmadjian, V. (2000) *Symbiosis. An Introduction to Biological Associations*. Oxford University Press, Oxford

Pascual-Garrido, A. (2011) *Insectivory of Nigerian Chimpanzees: Habitat Ecology and Harvesting Strategies*. PhD thesis. Universidad Complutense Madrid

Pascual-Garrido, A., Buba, U., Nodza, G. & Sommer, V. (2012) Obtaining raw material: plants as tool sources for Nigerian chimpanzees. *Folia Primatologica* 83, 24–44

Pebsworth, P.A., Hillier, S., Wendler, R., Glahn, R., Ta, C.A.K., Arnason, J.T. & Young, S.L. (2019) Geophagy among East African chimpanzees: consumed soils provide protection from plant secondary compounds and bioavailable iron. *Environmental Geochemistry and Health* 41, 2911–2927

Peltola, T. & Heikkilä, J. (2018) Outlaws or protected? DNA, hybrids, and biopolitics in a Finnish wolf-poaching case. *Society and Animals* 26, 197–216

PÉREZ-SALICRUP, D.R. & BARKER, M.G. (2000) Effect of liana cutting on water potential and growth of adult *Senna multijuga* (Caesalpinioideae) trees in a Bolivian tropical forest. *Oecologia* 124, 469–475

PERRY, S., BAKER, M., FEDIGAN, L., GROS-LOUIS, J., JACK, K., MACKINNON, K.C., et al. (2003) Social conventions in wild white-faced capuchin monkeys: evidence for traditions in a neotropical primate. *Current Anthropology* 44, 241–268

PETERSON, D. & GOODALL, J. (1993) *Visions of Caliban: On Chimpanzees and People.* Houghton Mifflin, Boston MA

PETERSON, D. & AMMANN, K. (2003) *Eating Apes.* University of California Press, Berkeley CA

PETRÁŠOVÁ, J., MODRÝ, D., HUFFMAN, M.A., MAPUA, M.I., BOBÁKOVÁ, L., MAZOCH, V., et al. (2010) Gastrointestinal parasites of indigenous and introduced primate species of Rubondo Island National Park, Tanzania. *International Journal of Primatology* 31, 920–936

PETRÁŠOVÁ, J., UZLÍKOVÁ, M., KOSTKA, M., PETRŽELKOVÁ, K.J., HUFFMAN, M.A. & MODRÝ, D. (2011) Diversity and host specificity of Blastocystis in syntopic primates on Rubondo Island, Tanzania. *International Journal for Parasitology* 41, 1113–1120

PETRŽELKOVÁ, K.J., HASEGAWA, H., MOSCOVICE, L.R., KAUR, T., ISSA, M. & HUFFMAN, M.A. (2006) Parasitic nematodes in the chimpanzee population on Rubondo Island, Tanzania. *International Journal of Primatology* 27, 767–778

PETRŽELKOVÁ, K.J., MODRÝ, D., POMAJBÍKOVÁ, K, PROFOUSOVÁ, I., SCHOVANCOVÁ, K., KAMLER, J. & JANOVÁ, E. (2008a) Factors influencing populations of entodinimorph ciliates in wild and captive chimpanzees. Abstracts, *XXII.* Congress, *International Primatological Society*, Edinburgh 03–08 Aug 08

PETRŽELKOVÁ, K.J., BOBÁKOVÁ, L, MOSCOVICE, L.A., MAPUA, M.I., HASEGAWA, H., HUFFMAN, M.A., PETRÁŠOVÁ, J. & KAUR, T. (2008b) Successful introduction of zoo chimpanzees onto Rubondo Island (Tanzania). Poster, 25th EAZA conference, Anwerp, Belgium, 16–20 Sep 08

PETRŽELKOVÁ, K.J., BOBÁKOVÁ, L. & MAZOCH, V. (2010a) *Rubondo Island and Mahale Mountains Chimpanzee Initiative: Integrating Socio-Ecological Research and Conservation Medicine.* Unpublished report to TAWIRI. Unpublished Report

PETRŽELKOVÁ, K.J., HASEGAWA, H., APPLETON, C.C., HUFFMAN, M.A., ARCHER, C.E., MOSCOVICE, L.R., et al. (2010b) Gastrointestinal parasites of the chimpanzee population introduced onto Rubondo Island National Park, Tanzania. *American Journal of Primatology* 72, 307–316

PIANKA, E.R. (1970) On r- and K-selection. *The American Naturalist* 104, 592–597

PIEL, A.K., STRAMPELLI, P., GREATHEAD, E., HERNANDEZ-AGUILAR, R.A., MOORE, J. & STEWART, F.A. (2017) The diet of open-habitat chimpanzees (*Pan troglodytes schweinfurthii*) in the Issa valley, western Tanzania. *Journal of Human Evolution* 112, 57–69

POMAJBÍKOVÁ, K., PETRŽELKOVÁ, K.J., PROFOUSOVÁ, I., PETRÁŠOVÁ, J., KIŠIDAYOVÁ, S., VARÁDYOVÁ, Z. & MODRÝ, D. (2010) A survey of entodiniomorphid ciliates in chimpanzees and bonobos. *American Journal of Physical Anthropology* 142, 42–48

POMAJBÍKOVÁ, K., PETRŽELKOVÁ, K.J., PETRÁŠOVÁ, J., PROFOUSOVÁ, I., KALOUSOVÁ, B., JIRKŮ, M., SÁ, R.M. & MODRÝ, D. (2012) Distribution of the entodiniomorphid ciliate *Troglocorys cava* Tokiwa, Modrý, Ito, Pomajbíková, Petrželková, &

Imai 2010, (Entodiniomorphida: Blepharocorythidae) in wild and captive chimpanzees. *Journal of Eukaryotic Microbiology* 59, 97–99

PRINGLE, R.M. (2005) The origins of the nile perch in Lake Victoria. *BioScience* 55, 780–787

PROFOUSOVÁ, I., MIHALIKOVÁ, K., LAHO, T., VÁRADYOVÁ, Z., PETRŽELKOVÁ, K.J., MODRÝ, D. & KIŠIDAYOVÁ, S. (2011) The ciliate, *Troglodytella abrassarti*, contributes to polysaccharide hydrolytic activities in the chimpanzee colon. *Folia Microbiologica* 56, 339–343

PRUETZ, J. (2006) Feeding ecology of savanna chimpanzees (*Pan troglodytes verus*) at Fongoli, Senegal. In *The Feeding Ecology of Great Apes and Other Primates* (eds G. HOHMANN, M.M. ROBBINS & C. BOESCH), pp. 161–182. Cambridge University Press, Cambridge

PRUETZ, J.D., FULTON, S.J., MARCHANT, L.F., MCGREW, W.C., SCHIEL, M. & WALLER, M. (2008) Arboreal nesting as anti-predator adaptation by savanna chimpanzees (*Pan troglodytes verus*) in Southeastern Senegal. *American Journal of Primatology* 70, 393–401

PRUETZ, J. & BERTOLANI, P. (2009) Chimpanzee (*Pan troglodytes verus*) behavioral responses to stresses associated with living in a savannah-mosaic environment: implications for Hominin adaptations to open habitats. *Paleoanthropology* 10, 252–262

PUSEY, A. (1998) *Rubondo Chimp Project*. Unpublished report

PUSEY, A., WILLIAMS, J. & GOODALL, J. (1997) The influence of dominance rank on the reproductive success of female chimpanzees. *Science* 277, 828–831

QUAMMEN, D. (2019) "I am scared all the time": chimps and people are clashing in rural Uganda. *National Geographic* (08 Nov) [accessed 12 Apr 21]

QUATTROCCHI, U. (2012) *CRC World Dictionary of Medicinal and Poisonous Plants: Common Names, Scientific Names, Eponyms, Synonyms, and Etymology*. CRC Press, Boca Raton FL

RADHAKRISHNA, S. (2017) Cultural and religious aspects of primate conservation. In *The International Encyclopedia of Primatology* (ed A. FUENTES), pp. 1–8. Wiley-Blackwell, Hoboken NJ

RALLS, K., FRANKHAM, R. & BALLOU, J.D. (2013) Inbreeding and outbreeding. In *Encyclopedia of Biodiversity* (2nd ed.) (ed S.A. LEVIN), pp. 245–252. Elsevier, Amsterdam

REED, D.L., LIGHT, J.E., ALLEN, J. & KIRCHMAN, J. (2007) Pair of lice lost or parasites regained: the evolutionary history of Anthropoid primate lice. *BMC Biology* 5, 7

REEMTSMA, K. (1995) Xenotransplantation: a historical perspective. *ILAR Journal* 37, 9–12

REYNOLDS, V. (2005) *The Chimpanzees of the Budongo Forest: Ecology, Behaviour and Conservation*. Oxford University Press, Oxford

ROBINSON, J. & ROBINSON, P. (1998) *Chimpanzee Status Survey and Habituation Project: Progress Report 1998*. Frankfurt Zoological Society. Unpublished Report

RODGERS, W.A., LUDANGA, R.I. & DESUZO, H.P. (1977) Biharamulo, Burigi, and Rubondo Island Game Reserves. *Tanzania Notes Records* 81/82, 99–124

ROSE, A.L., MITTERMEIER, R.A., LANGRAND, O., AMPADU-AGYEI, O., BUTYNSKI, T.M. & AMMANN, K. (2003) *Consuming Nature. A Photo Essay on African Rain Forest Exploitation*. Altisima Press, Palos Verdes CA

ROUND, J.L. & MAZMANIAN, S.K. (2009) The gut microbiota shapes intestinal immune responses during health and disease. *Nature Reviews Immunology* 9, 313–323

RUTHERFORD, S. (2018) The Anthropocene's animal? Coywolves as feral cotravelers. *Environment and Planning E: Nature and Space* 1, 206–223

VAN SCHAIK, C.P., TERBORGH, J.W. & WRIGHT, S.J. (1993) The phenology of tropical forests: adaptive significance and consequences for primary consumers. *Annual Review of Ecology and Systematics* 24, 353–377

SCHMIDT, J.M., DE MANUEL, M., MARQUES-BONET, T., CASTELLANO, S. & ANDRÉS, A.M. (2019) The impact of genetic adaptation on chimpanzee subspecies differentiation. *PLOS Genetics* 15, e1008485

SCHMIDT, R. (1898) *Deutschlands Kolonien. Vol. 1. Verlag des Vereins der Bücherfreunde.* Schall und Grund, Berlin

SCHNEIDER, G. (2017) Der Mister Serengeti vom Zürichsee. (Interview with Markus Borner). *Zürichsee-Zeitung* (29 Jan) [accessed 30 Dec 20]

SCHNITZER, S.A. & CARSON, W.P. (2001) Treefall gaps and the maintenance of species diversity in a tropical forest. *Ecology* 82, 913–919

SCHOENE C.U.R. & BREND S.A. (2002) Primate sanctuaries: a delicate conservation approach. *South African Journal of Wildlife Research* 32, 109–113

SCHÜRMANN, FELIX (2016) Rubondo und eine Reise dorthin. Der Feldaufenthalt in der Geschichtswissenschaft – und unter afrikanischen Wildtieren. In *Den Fährten folgen* (ed Forschungsschwerpunkt "Tier – Mensch – Gesellschaft". Methoden interdisziplinärer Tierforschung), pp. 133–154. Transcript, Bielefeld

SCHÜRMANN, F. (2017a) Grzimeks Afrika. ... zwischen westlichem Naturschutzkonzept und kolonialen Klischees. *Zeitgeschichte-online* (13 Mar) [dev.zeitgeschichte-online.de/film/grzimeks-afrika; accessed 08 Jun 20]

SCHÜRMANN, FELIX (2017b) Heimkehr ins Neuland. Die erste Auswilderung von Schimpansen und ihre Kontexte im postkolonialen Tansania, 1965–1966. In *Vielfältig verflochten* (ed Forschungsschwerpunkt "Tier – Mensch – Gesellschaft". Methoden interdisziplinärer Tierforschung), pp. 275–292. Transcript, Bielefeld

SCHÜRMANN, F. (2021) Naturverinselung im Viktoriasee: Ökologische Erbschaften der transimperialen Kampagne gegen die Schlafkrankheit. *Transimperial History Blog*, (23 Jul) [www.transimperialhistory.com; accessed 11 Aug 21]

SCHÜSSLER, D., BLANCO, M.B., SALMONA, J., POELSTRA, J., ANDRIAMBELOSON, J.B., MILLER, A., et al. (2020) Ecology and morphology of mouse lemurs (*Microcebus spp.*) in a hotspot of microendemism in northeastern Madagascar, with the description of a new species. *American Journal of Primatology* 82, e23180

SEDDON, J.M. & SCHULTZ, B. (2020) Koala conservation in Queensland, Australia: a role for assisted gene flow for genetic rescue? in *Conservation Genetics in Mammals: Integrative Research Using Novel Approaches* (eds J. ORTEGA & J.E. MALDONADO), pp. 331–349. Springer, Cham

SEDDON, P.J., ARMSTRONG, D.P. & MALONEY, R.F. (2007) Developing the science of reintroduction biology. *Conservation Biology* 21, 303–312

SEWIG, C. (2009) *Grzimek: Der Mann, der die Tiere liebte.* Bastei-Lübbe, Bergisch Gladbach

SHEA, B.T. (1984) An allometric perspective on the morphological and evolutionary relationships between pygmy (*Pan paniscus*) and common (*Pan troglodytes*) chimpanzees. In *The Pygmy Chimpanzee: Evolutionary Biology and Behavior* (ed R.L. SUSMAN), pp. 89–130. Springer, Boston MA

SHUMAKER, R.W., WALKUP, K.R. & BECK, B.B. (2011) *Animal Tool Behavior: The Use and Manufacture of Tools by Animals.* Johns Hopkins University Press, Baltimore MD

SILK, J.B. (2014) The evolutionary roots of lethal conflict. *Nature* 513, 321–322

SMITH-RAMIREZ, C. & ARMESTO, J.J. (1994) Flowering and fruiting patterns in the temperate rainforest of Chiloe, Chile–ecologies and climatic constraints. *Journal of Ecology* 82, 353–365

SOMMER, V. (2021) Artenschutz im Zoo – ein Etikettenschwindel. *APuZ – Aus Politik und Zeitgeschichte* 9, 35–38

SOMMER, V. & PARISH, A.R. (2010) Living Differences. In *Homo Novus – A Human Without Illusions* (eds U.J. FREY, C. STÖRMER & K.P. WILLFÜHR), pp. 19–33. Springer, Berlin/Heidelberg

SOMMER, V. & ROSS, C. (eds) (2011) *Primates of Gashaka. Socioecology and Conservation in Nigeria's Biodiversity Hotspot. Developments in Primatology: Progress and Prospects 35.* Springer, New York NY

SOMMER, V. & ROSS, C. (2011) Exploring and protecting West Africa's primates: The Gashaka Primate Project in context. In *Primates of Gashaka: Socioecology and Conservation in Nigeria's Biodiversity Hotspot* (eds V. SOMMER & C. ROSS), pp. 1–37. Springer, New York NY

SOMMER, V., ADANU, J., FAUCHER, I. & FOWLER, A. (2004) Nigerian chimpanzees (*Pan troglodytes vellerosus*) at Gashaka: two years of habituation efforts. *Folia Primatologica* 75, 295–316

SOMMER, V., BAUER, J., FOWLER, A. & ORTMANN, S. (2011) Patriarchal chimpanzees, matriarchal bonobos: potential ecological causes of a *Pan* dichotomy. In *Primates of Gashaka: Socioecology and Conservation in Nigeria's Biodiversity Hotspot* (eds V. SOMMER & C. ROSS), pp. 469–501. Springer, New York NY

SOMMER, V., LOWE, A., JESUS, G., ALBERTS, N., BOUQUET, Y., INGLIS, D.M., et al. (2016) Antelope predation by Nigerian forest baboons: ecological and behavioural correlates. *Folia Primatologica* 87, 67–90

SOMMER, V., BUBA, U., JESUS, G. & PASCUAL-GARRIDO, A. (2017) Sustained myrmecophagy in Nigerian chimpanzees: preferred or fallback food? *American Journal of Physical Anthropology* 162, 328–336

SONG, S.J., LAUBER, C., COSTELLO, E.K., LOZUPONE, C.A., HUMPHREY, G., BERG-LYONS, D., et al. (2013) Cohabiting family members share microbiota with one another and with their dogs. *eLife* 2, e00458

SPEITKAMP, W. (2005) *Deutsche Kolonialgeschichte.* Reclam, Stuttgart

SPIEGEL, M. (1988) *The Dreaded Comparison: Human and Animal Slavery.* Heretic Books, London

STAGER, J.C., RYVES, D.B., CHASE, B.M. & PAUSATA, F.S.R. (2011) Catastrophic drought in the Afro-Asian monsoon region during Heinrich event 1. *Science* 331, 1299–1302

STANFORD, C.B., WALLIS, J., MPONGO, E. & GOODALL, J. (1994) Hunting decisions in wild chimpanzees. *Behaviour* 131, 1–18

STANFORD, C.B. & O'MALLEY, R.C. (2008) Sleeping tree choice by Bwindi chimpanzees. *American Journal of Primatology* 70, 642–649

STEVENS, G.C. (1987) Lianas as structural parasites: the *Bursera simaruba* example. *Ecology* 68, 77–81

STEVENSON, T. & FANSHAWE, J. (2020) *Field Guide to the Birds of East Africa*: Kenya, Tanzania, Uganda, Rwanda, Burundi. Princeton University Press, Princeton NY

STEWART, F.A. (2011) The *Evolution of Shelter: Ecology and Ethology of Chimpanzee Nest Building.* PhD thesis. University of Cambridge

STILES, D., REDMOND, I., CRESS, D., NELLEMANN, C. & FORMO, R.K. (2013) *Stolen Apes – the Illicit Trade in Chimpanzees, Gorillas, Bonobos and Orangutans: A Rapid Response Asssessment*. United Nations Environment Programme, GRID-Arendal

STRIER, K.B. (2017) What does variation in primate behavior mean? *American Journal of Physical Anthropology* 162, 4–14

SUGIYAMA, Y. (2004) Demographic parameters and life history of chimpanzees at Bossou, Guinea. *American Journal of Physical Anthropology* 124, 154–165

TAABU-MUNYAHO, A., MARSHALL, B., TOMASSON, T. & MARTEINSDOTTIR, G. (2016) Nile perch and the transformation of Lake Victoria. *African Journal of Aquatic Science* 41, 127–142

Tanzania National Parks (TANAPA) (2003) *General Management Plan for Rubondo Island National Park* 2003–2013. Arusha, Tanzania

Tanzania National Parks (TANAPA) (2020) *Visitors. Tanzania National Parks.* [www.tanzaniaparks.go.tz/uploads/publications/en-1568717749-TEN%20YEARS%20ARRIVAL%20TRENDS.pdf; accessed 12 May 21]

Tanzania Wildlife Research Institute (TAWIRI) (2018) *Tanzania Chimpanzee Conservation Action Plan 2018–2023*. Arusha, Tanzania

TELEKI, G. (1989) Population status of wild chimpanzees (*Pan troglodytes*) and threats to survival. In *Understanding Chimpanzees* (eds P.G. HELTNE & L.A. MARQUARDT), pp. 312–353. Harvard University Press, Cambridge MA

TELEKI, G. (2001) Sanctuaries for ape refugees. In *Great Apes and Humans: The Ethics of Coexistence* (eds B.B. BECK, T. STOINSKI, H. HUTCHIN, T.L. MAPLE, B. NORTON, A. ROWAN, et al.), pp. 133–149. Smithsonian Institution Press, Washington DC

TRICONE, F. (2018) Assessment of releases of translocated and rehabilitated Yucatán black howler monkeys (*Alouatta pigra*) in Belize to determine factors influencing survivorship. *Primates* 59, 69–77

TRYON, C.A., FAITH, J.T., PEPPE, D.J., BEVERLY, E.J., BLEGEN, N., BLUMENTHAL, S.A., et al. (2016) The Pleistocene prehistory of the Lake Victoria basin. The *African Quaternary: Environments, Ecology and Humans Inaugural AFQUA Conference*, 404, 100–114

TSUKAYAMA, P., BOOLCHANDANI, M., PATEL, S., PEHRSSON, E.C., GIBSON, M.K., CHIOU, K.L., et al. (2018) Characterization of wild and captive baboon gut microbiota and their antibiotic resistomes. *mSystems* 3, e00016-18

TURNER, I.M. (2001) *The Ecology of Trees in the Tropical Rain Forest*. Cambridge University Press, Cambridge

TUTIN, C.E.G. & FERNANDEZ, M. (1984) Nationwide census of gorilla (*Gorilla g. gorilla*) and chimpanzee (*Pan t. troglodytes*) populations in Gabon. *American Journal of Primatology* 6, 313–336

TUTIN, C.E.G. & FERNANDEZ, M. (1993) Relationships between minimum temperature and fruit production in some tropical forest trees in Gabon. *Journal of Tropical Ecology* 9, 241–248

TWEHEYO, M. & BABWETEERA, F. (2007) Production, seasonality and management of chimpanzee food trees in Budongo Forest, Uganda. *African Journal of Ecology* 45, 535–544

United Nations Population Fund (UNFPA) (2012) *Tanzania: A Youthful and Rapidly Growing Nation*. United Nations Population Fund

VALLO, P., PETRŽELKOVÁ, K.J., PROFOUSOVÁ, I., PETRÁŠOVÁ, J., POMAJBÍKOVÁ, K., LEENDERTZ, F., et al. (2012) Molecular diversity of entodiniomorphid ciliate

Troglodytella abrassarti and its coevolution with chimpanzees. *American Journal of Physical Anthropology* 148, 525–533

VAN LAWICK-GOODALL, J. (1968) The behaviour of free-living chimpanzees in the Gombe Stream Reserve. *Animal Behaviour Monographs* 1, 161–311

VIDEAN, E.N. (2006) Bed-building in captive chimpanzees (*Pan troglodytes*): the importance of early rearing. *American Journal of Primatology* 68, 745–751

VIGGERS, K., LINDENMAYER, D. & SPRATT, D. (1993) The importance of disease in reintroduction programmes. *Wildlife Research* 20, 687–698

WAKEFIELD, M.L. (2008) Grouping patterns and competition among female *Pan troglodytes schweinfurthii* at Ngogo, Kibale National Park, Uganda. *International Journal of Primatology* 29, 907–929

WALKER, J.A., JORDAN, V.E., STORER, J.M., STEELY, C.J., GONZALEZ-QUIROGA, P., BECKSTROM, T.O., et al. (2019) Alu insertion polymorphisms shared by *Papio* baboons and *Theropithecus gelada* reveal an intertwined common ancestry. *Mobile DNA* 10, 46–58

WALKER, K.K., WALKER, C.S., GOODALL, J. & PUSEY, A.E. (2018) Maturation is prolonged and variable in female chimpanzees. *Journal of Human Evolution* 114, 131–140

WALTER, J. & O'MAHONY, L. (2019) The importance of social networks: an ecological and evolutionary framework to explain the role of microbes in the aetiology of allergy and asthma. *Allergy* 74, 2248–2251

WATTS, D.P., POTTS, K.B., LWANGA, J.S. & MITANI, J.C. (2012) Diet of chimpanzees (*Pan troglodytes schweinfurthii*) at Ngogo, Kibale National Park, Uganda, 2. Temporal variation and fallback foods. *American Journal of Primatology* 74, 130–144

WĘCEK, K., HARTMANN, S., PAIJMANS, J.L.A., TARON, U., XENIKOUDAKIS, G., CAHILL, J.A., et al. (2017) Complex admixture preceded and followed the extinction of wisent in the wild. *Molecular Biology and Evolution* 34, 598–612

WHITE, K.L., EASON, D.K., JAMIESON, I.G. & ROBERTSON, B.C. (2015) Evidence of inbreeding depression in the critically endangered parrot, the kakapo. *Animal Conservation* 18, 341–347

WHITELEY, A.R., FITZPATRICK, S.W., FUNK, W.C. & TALLMON, D.A. (2015) Genetic rescue to the rescue. *Trends in Ecology and Evolution* 30, 42–49

WHITEN, A., GOODALL, J., McGREW, W.C., NISHIDA, T., REYNOLDS, V., SUGIYAMA, Y., et al. (1999) Cultures in chimpanzees. *Nature* 399, 682–685

WHITEN, A. & McGREW, W.C. (2001) Piecing together the history of our knowledge of chimpanzee tool use. *Nature* 411, 413–413

WHITEN, A., HORNER, V. & MARSHALL-PESCINI, S. (2003) Cultural panthropology. *Evolutionary Anthropology* 12, 92–105

WILDMAN, D.E., UDDIN, M., LIU, G., GROSSMAN, L.I. & GOODMAN, M. (2003) Implications of natural selection in shaping 99.4% nonsynonymous DNA identity between humans and chimpanzees: enlarging genus Homo. *Proceedings of the National Academy of Sciences* 100, 7181–7188

WILLIAMS, C. (2020) *Evolution of the Primate Gut Microbiome.* PhD thesis. University College London

WILLIAMS, C., MSINDAI, N.J., SOMMER, V., PIEL, A., STEWART, F., CHATTERJEE, H. & SPRATT, D. (in prep.) Wild versus captive primate gut microbiota: Dramatic changes persist for half a century through multiple generations

WILLIAMS, J.M., PUSEY, A.E., CARLIS, J.V., FARM, B.P. & GOODALL, J. (2002) Female competition and male territorial behaviour influence female chimpanzees' ranging patterns. *Animal Behaviour* 63, 347–360

WILLIAMS, J.M., OEHLERT, G., CARLIS, J., & PUSEY, A.E. (2004) Why do male chimpanzees defend a group range? Reassessing male territoriality. *Animal Behaviour* 68, 523–532

WILSON, M.L., HAUSER, M.D. & WRANGHAM, R.W. (2007) Chimpanzees (*Pan troglodytes*) modify grouping and vocal behaviour in response to location-specific risk. *Behaviour* 144, 1621–1653

WILSON, M. (2013) Chimpanzees, warfare, and the invention of peace. In *War, Peace, and Human Nature: The Convergence of Evolutionary and Cultural Views* (ed D.P. FRY), pp. 361–388. Oxford University Press, Oxford

WILSON, M.L., BOESCH, C., FRUTH, B., FURUICHI, T., GILBY, I.C., HASHIMOTO, C., et al. (2014) Lethal aggression in *Pan* is better explained by adaptive strategies than human impacts. *Nature* 513, 414–417

WITHERS, M. & HOSKING, D. (2002) *Wildlife of East Africa*. Princeton University Press, Princeton NY

WOOD, B.M., WATTS, D.P., MITANI, J.C. & LANGERGRABER, K.E. (2017) Favorable ecological circumstances promote life expectancy in chimpanzees similar to that of human hunter-gatherers. *Journal of Human Evolution* 105, 41–56

WOOD, K.L., TENGER, B., MORF, N.V. & KRATZER, A. (2014) *Report to CITES: CITES-Listed Species at Risk from the Illegal Trafficking in Bushmeat. Results of A 2012 Study in Switzerland's International Airports*. Tengwood Organization, Switzerland. Unpublished Report to CITES

WORTON, B.J. (1989) Kernel methods for estimating the utilization distribution in home-range studies. *Ecology* 70, 164–168

WRANGHAM, R.W. (1977) Feeding behaviour of chimpanzees in Gombe National Park, Tanzania. In *Primate Ecology* (ed T.H. CLUTTON-BROCK), pp. 504–538. Academic Press, London

WRANGHAM, R.W. (2000) Why are male chimpanzees more gregarious than mothers? in *Primate Males: Causes and Consequences of Variation in Group Composition* (ed P.M. KAPPELER), pp. 248–258. Cambridge University Press, Cambridge

WRANGHAM, R.W. (2009) *Catching Fire: How Cooking Made Us Human*. Profile Books, London

WRANGHAM, R.W., McGREW, W.C., DE WAAL, F.B.M. & HELTNE, P.G. (eds) (1994) *Chimpanzee Cultures*. Harvard University Press, Cambridge MA

WRANGHAM, R.W. & SMUTS, B.B. (1980) Sex differences in the behavioural ecology of chimpanzees in the Gombe National Park, Tanzania. *Journal of Reproduction and Fertility* 28, 13–31

WRANGHAM, R.W. & PETERSON, D. (1996) *Demonic Males. Apes and the Origins of Human Violence*. Mariner Books, Boston MA

WROGEMANN, D. (1992) *Wild Chimpanzees in Lope*, Gabon: Census Method and Habitat Use. PhD thesis. Universität Bremen

YAMAGIWA, J. (1999) Socioecological factors influencing population structure of gorillas and chimpanzees. *Primates* 40, 87–104

YATSUNENKO, T., REY, F.E., MANARY, M.J., TREHAN, I., DOMINGUEZ-BELLO, M.G., CONTRERAS, M., et al. (2012) Human gut microbiome viewed across age and geography. *Nature* 486, 222–227

YERKES, R.M. & LEARNED, B.W. (1925) *Chimpanzee Intelligence and its Vocal Expressions*. Williams and Wilkins, Baltimore MD

YODER, A.D., WEISROCK, D.W., RASOLOARISON, R.M. & KAPPELER, P.M. (2016) Cheirogaleid diversity and evolution: big questions about small primates. In *The Dwarf and Mouse Lemurs of Madagascar: Biology, Behavior and Conservation*

Biogeography of the Cheirogaleidae (eds S.M. LEHMAN, U. RADESPIEL & E. ZIMMERMANN), pp. 3–20. Cambridge University Press, Cambridge

ZACHOS, F.E. (2016) *Species Concepts in Biology: Historical Development, Theoretical Foundations and Practical Relevance.* Springer, Cham

ZINNER, D., ARNOLD, M.L. & ROOS, C. (2011) The strange blood: natural hybridization in primates. *Evolutionary Anthropology* 20, 96–103

ZINNER, D. & ROOS, C. (2016) Primate taxonomy and conservation. In *Ethnoprimatology: Primate Conservation in the 21st Century* (ed M.T. WALLER), pp. 193–213. Springer, Cham

Appendix: Publications about Rubondo Island, Its History and Wildlife

The compilations are presented in chronological order to provide a timeline of how certain topics unfold

Contributions in Journals, Edited Volumes, Books and Research Repositories

McCulloch, Brian; Peter L. Achard (1965) Rhino reserve in Lake Victoria. *Oryx* 8, 162–163

Grzimek, Bernhard (1966) Apes travel from Europe to Africa. *African Wildlife* 20, 271–288

Grzimek, Bernhard (1969) Menschenaffen reisten von Europa nach Afrika. In *Grzimek unter Afrikas Tieren: Erlebnisse, Beobachtungen, Forschungsergebnisse* (ed B Grzimek), pp. 11–36. Ullstein, Berlin/Frankfurt a.M.

McCulloch, Brian; Peter L. Achard (1969) Mortalities associated with the capture, translocation, trade and exhibition of black rhinoceroses (*Diceros bicornis*). *International Zoo Yearbook* 9, 184–191

Grzimek, Bernhard (1971) How Europe exported apes to Africa. In *Among Animals of Africa* (ed B Grzimek), pp. 11–37. Harper Collins, London (USA edition 1970, Stein and Day, New York NY)

Grzimek, Bernhard (1974) *Auf den Mensch gekommen: Erfahrungen mit Leuten.* Bertelsmann, Munich/Gütersloh/Vienna [p. 464]

Anzenberger, Gustl (1977) Ethological study of African carpenter bees of the genus *Xylocopa* (Hymenoptera, Anthophoridae). *Zeitschrift für Tierpsychologie* 44, 337–374

Rodgers, W.A.; R.I. Ludanga; H.P. De Suzo (1977) An ecological survey of Biharamulo, Burigi and Rubondo Island Game Reserves. *Tanzania Notes and Records* 81, 99–124

Rodgers, W.A. (1981) The distribution and conservation status of colobus monkeys in Tanzania. *Primates* 22, 33–45

Borner, Monica (1985) The rehabilitated chimpanzees of Rubondo Island. *Oryx* 19, 151–154

Anzenberger, Gustl (1986) How do carpenter bees recognize the entrance of their nests? An experimental investigation in a natural habitat. *Ethology* 71, 54–62

Borner, Monica (1988) Translocation of 7 mammal species to Rubondo Island National Park in Tanzania. In *Translocation of Wild animals* (eds L. Nielsen

& R.D. Brown), pp. 117–122. Wisconsin Humane Society and Caesar Kleberg Wildlife Research Institute, Milwaukee WI

Grzimek, Bernhard (1988) Die Auswilderung von Schimpansen. In *Grzimeks Enzyklopädie Säugetiere Vol. 2* (ed Bernhard Grzimek), pp. 482–485. Kindler, Munich

Kruuk, H.; P.C. Goudswaard (1990) Effects of changes in fish populations in Lake Victoria on the food of otters (*Lutra maculicollis Schinz* and *Aonyx capensis Lichtenstein*). *African Journal of Ecology* 28, 322–329

Müller, Guido; Gustl Anzenberger (1995) *Chimpanzees (Pan troglodytes) of Rubondo Island, Tanzania. Anthropologisches Institut, der Universität Zürich.* (Pilot study, German with English summary) doi.org/10.13140/RG.2.2.31713 .12648 [www.researchgate.net; posted online in 2020]

Matsumoto–Oda, Akiko (2000) Chimpanzees in the Rubondo Island National Park, Tanzania. *Pan Africa News* 7, 16–17

Mnaya, Bakari; Eric Wolanski (2002) Water circulation and fish larvae recruitment in papyrus wetlands, Rubondo Island, Lake Victoria. *Wetlands Ecology and Management* 10, 131–141

Hasegawa, Hideo; Yatsukaho Ikeda; Akiko Fujisaki; Liza R. Moscovice; Klára J. Petrželková; Taranjit Kaur; Michael A. Huffman (2005) Morphology of chimpanzee pinworms Enterobius (*Enterobius*) *anthropopitheci* (Gedoelst 1916) (Nematoda: *Oxyuridae*) collected from chimpanzees *Pan troglodytes* on Rubondo Island, Tanzania. *Journal of Parasitolology* 91, 1314–1317

Petrželková, Klára J.; Hideo Hasegawa; Liza R. Moscovice; Taranjit Kaur; Mwanahamisi I. Mapua; Michael A. Huffman (2006) Parasitic nematodes in the chimpanzee population on Rubondo island, Tanzania. *International Journal of Primatology* 27, 767–777

Moscovice, Liza R.; Mwanahamisi I. Mapua; Klára J. Petrželková; Nicholas S. Keuler; Charles T. Snowdon; Michael A. Huffman (2007) Fruit availability, chimpanzee diet, and grouping patterns on Rubondo Island, Tanzania. *American Journal of Primatology* 69, 487–502

Huffman, Michael A.; Klára J. Petrželková; Liza R. Moscovice; Mwanahamisi I. Mapua; Lucia Bobáková; Vladimír Mazoch; Jatinder Singh; Taranjit Kaur (2008) Introduction of chimpanzees onto Rubondo Island National Park, Tanzania. *(SSC Re-Introduction Specialist Group, Abu Dhabi) Re-introduction NEWS (Special Issue)* 27, 213–216

Moscovice, Liza R.; Frank Mbago; Charles T. Snowdon; Michael A. Huffman (2010) Ecological features and ranging patterns at a chimpanzee release site on Rubondo Island, Tanzania. *Biological Conservation* 143, 2711–2721

Petrášová, Jana; David Modrý; Michael A. Huffman; Mwanahamisi I. Mapua; Lucia Bobáková; Vladimír Mazoch; Jatinder Singh; Taranjit Kaur; Klara J. Petrželková (2010) Gastrointestinal parasites of indigenous and introduced primate species of Rubondo Island National Park, Tanzania. *International Journal of Primatology* 31, 920–936

Petrželková, Klára J.; Hideo Hasegawa; Chris C. Appleton; Michael A. Huffman; Colleen E. Archer; Liza R. Moscovice; Mwanahamisi I. Mapua; Jatinder Singh; Taranjit Kaur (2010) Gastrointestinal parasites of the chimpanzee population introduced onto Rubondo Island National Park, Tanzania. *American Journal of Primatology* 72, 307–316

POMAJBÍKOVÁ, KATEŘINA; KLÁRA J. PETRŽELKOVÁ; ILONA PROFOUSOVÁ; JANA PETRÁŠOVÁ; DAVID MODRÝ (2010) Discrepancies in the occurrence of Balantidium coli between wild and captive African great apes. *Journal of Parasitology* 96, 1139–1144

POMAJBÍKOVÁ, KATEŘINA; KLÁRA J. PETRŽELKOVÁ; ILONA PROFOUSOVÁ; JANA PETRÁŠOVÁ; SVETLANA KIŠIDAYOVÁ; ZORA VARÁDYOVÁ; DAVID MODRÝ (2010) A survey of entodiniomorphid ciliates in chimpanzees and bonobos. *American Journal of Biological Anthropology* 142, 42–48

REED-SMITH, JANICE; ISAAC OLUOCH; MARK ORIGA; TUA SAMWELI KIHEDU; MAJALIWA M. MUHABI; MANDY M. YUSUF; MORDECAI OGADA; ALEX LOBORA; THOMAS SERFASS (2010) Consumptive uses of and lore pertaining to spotted-necked otters in East Africa: a preliminary report from the Lake Victoria area of Kenya, Tanzania, and Uganda. *IUCN/SCC Otter Specialist Group Bulletin* 27, 85–88

PETRÁŠOVÁ, JANA M.; M. UZLÍKOVÁ; M. KOSTKA; KLÁRA J. PETRŽELKOVÁ; MICHAEL A. HUFFMAN; DAVID MODRÝ (2011) Diversity and host specificity of blastocystis in syntopic primates on Rubondo island, Tanzania. *International Journal for Parasitology* 41, 1113–1120

POMAJBÍKOVÁ, KATEŘINA; KLÁRA J. PETRŽELKOVÁ; JANA PETRÁŠOVÁ; ILONA PROFOUSOVÁ; BARBORA KALOUSOVÁ; MILOSLAV JIRKŮ; RUI M. SÁ; DAVID MODRÝ (2012) Distribution of the Entodiniomorphid Ciliate *Troglocorys cava* Tokiwa, Modrý, Ito, Pomajbíková, Petrželková, and Imai (Entodiniomorphida: Blepharocorythidae) in wild and captive chimpanzees. *Journal of Eukaryotic Microbiology* 59, 97–99

AMULIKE, BRIDGET; SADIE S. STEVENS; THOMAS L. SERFASS (2013) Enhancing tourist opportunities to view spotted-necked otters (*Lutra maculicollis*) at Rubondo Island National Park: can the *apriori* location of latrines simplify identifying best viewing areas? *African Journal of Ecology* 51, 609–617

MWAMBOLA, SIMON; JASPER LJUMBA; WICKSON KIBASA; EMANUEL MASENGA; ERNEST EBLATE; CANISIUS J. KAYOMBO (2014) Feeding preference of the African elephant (*Loxodanta africana*) on woody plant species in Rubondo Island National Park (RINP), Tanzania. *American Journal of Research Communication* 2, 102–113

REED-SMITH, JANICE; THOMAS SERFASS; TUA S. KIHEDU; MAJALIWA MUSSA (2014) Preliminary report on the behavior of spotted-necked otter (*Lutra maculicollis*, Lichtenstein, 1835) living in a lentic ecosystem. *Zoo Biology* 33, 121–130

DOLEZALOVA, JANA; MIROSLAV OBORNIK; EVA HAJDUSKOVA; MILAN JIRKU; KLÁRA J. PETRŽELKOVÁ; PETRA BOLECHOVA; CRISTINA CUTILLAS; ROCIO CALLEJON; JAROS JOZEF; ZUZANA BERANKOVA; DAVID MODRÝ (2015) How many species of whipworms do we share? Whipworms from man and other primates form two phylogenetic lineages. *Folia Parasitologica (Praha)* 62, 63

MWAMBOLA, SIMON; JASPER IJUMBA; WICKSON KIBASA; EMMANUEL MASENGA; ERNEST EBLATE; LINUS MUNISHI (2016) Population size estimates and distribution of the African elephant using the dung surveys method in Rubondo Island National Park, Tanzania. *International Journal of Biodiversity and Conservation* 8, 113–119

SCHÜRMANN, FELIX (2016) Rubondo und eine Reise dorthin. Der Feldaufenthalt in der Geschichtswissenschaft – und unter afrikanischen Wildtieren. In *Den Fährten folgen* (ed Forschungsschwerpunkt 'Tier – Mensch – Gesellschaft'. Methoden interdisziplinärer Tierforschung), pp. 133–154. Transcript, Bielefeld

SCHÜRMANN, FELIX (2017) Heimkehr ins Neuland. Die erste Auswilderung von Schimpansen und ihre Kontexte im postkolonialen Tansania, 1965–1966. In

Vielfältig verflochten (ed Forschungsschwerpunkt 'Tier – Mensch – Gesellschaft'. Methoden interdisziplinärer Tierforschung), pp. 275–292. Transcript, Bielefeld

BECK, BENJAMIN B. (2018) *Unwitting Travelers: A History of Primate Reintroduction.* Salt Water Media, Berlin MD, pp. 11–12

REED-SMITH, JANICE; DAVID ROWE-ROWE; J. JACQUES; MICHAEL SOMERS (2018) Spotted-necked otters. In *The Global Otter Conservation Strategy* (eds N. DUPLAIX & M. SAVAGE), pp. 102–109. IUCN/SSC Otter Specialist Group and Four Corners Institute, Salem OR

MSINDAI, JOSEPHINE N.; CHRISTIAN ROOS; FELIX SCHÜRMANN; VOLKER SOMMER (2021) Population history of chimpanzees introduced to Lake Victoria's Rubondo Island. *Primates* 62, 253–265

SCHÜRMANN, FELIX (2021) Naturverinselung im Viktoriasee: Ökologische Erbschaften der transimperialen Kampagne gegen die Schlafkrankheit. *Transimperial History Blog* (23 July) [www.transimperialhistory.com, accessed 11 August 2021]

WILLIAMS, C.; MSINDAI, N.J.; SOMMER, V.; PIEL, A.; STEWART, F.; CHATTERJEE, H.; SPRATT, D. (in prep.) Wild versus captive primate gut microbiota: Dramatic changes persist for half a century through multiple generations

Abstracts and Conference Presentations

MOSCOVICE, LIZA; MICHAEL A. HUFFMAN (2001) Ecology and ranging habits of an introduced group of chimpanzees on Rubondo Island National Park, Tanzania – a preliminary report. Poster, *4th International Saga Symposium*, Okayama, Japan, 15–17 Nov 01

MOSCOVICE, LIZA R.; KLÁRA J. PETRŽELKOVÁ; MWANAHAMISI I. MAPUA; MICHAEL A. HUFFMAN; CHARLES T. SNOWDON; FRANK MBAGO; TARANJIT KAUR; JATINDER SINGH; GIULIA GRAZIANI (2004) Role of lianas for introduced chimpanzees (*Pan troglodytes*) on Rubondo Island Tanzania. *Folia Primatologica* 75, 933

MOSCOVICE, LIZA R. (2006) Behavioral ecology of chimpanzees (*Pan troglodytes*) on Rubondo Island, Tanzania: Habitat, diet, grouping and ranging at a release site. *Dissertation Abstracts International* B 67: AADAA

BOBÁKOVÁ, LUCIA; KLÁRA J. PETRŽELKOVÁ; LIZA R. MOSCOVICE; MWANAHAMISI I. MAPUA; HEIDO HASEGAWA; MICHAEL A. HUFFMAN; JANA PETRÁŠOVÁ; TARANJIT KAUR (2007) Ecology of an introduced chimpanzee population, Rubondo Island (Tanzania). Poster, *Proceedings of 6th International Zoo and Wildlife Research Conference on Behaviour, Physiology and Genetics*, Berlin, Germany, 07–10 Oct 07

STEVENS, SADIE; THOMAS L. SERFASS; JOHN F. ORGAN (2007) Otters and wildlife tourism: A recipe for conservation success? Oral presentation, *Proceedings of the 5th TAWIRI Conference*, Arusha, Tanzania, 03–06 Dec 07

MAPUA, MWANAHAMISI I.; LUCIA BOBÁKOVÁ; KLÁRA J. PETRŽELKOVÁ; LIZA R. MOSCOVICE; HIDEO HASEGAWA; MICHAEL A. HUFFMAN; JANA PETRÁŠOVÁ; TARANJIT KAUR (2008) Long-term research of the chimpanzee population released onto Rubondo Island NP (Tanzania). Poster, *Zoological Days*, Brno, Czech Republic, 14–15 Feb 08

MAZOCH, VLADIMÍR; KLÁRA J. PETRŽELKOVÁ; LUCIA BOBÁKOVÁ; MWANAHAMISI I. MAPUA; MICHAEL A. HUFFMAN; TARANJIT KAUR (2008) Nesting behavior of introduced chimpanzees of Rubondo Island NP (Tanzania). Poster, *Zoological Days*, Brno, Czech Republic, 14–15 Feb 08

PETRŽELKOVÁ, KLÁRA J.; DAVID MODRÝ; KATEŘINA POMAJBÍKOVÁ; LONA PROFOUSOVÁ; KATEŘINA SCHOVANCOVÁ; JIRI KAMLER; EVA JANOVÁ (2008) Factors influencing populations of entodinimorph ciliates in wild and captive chimpanzees. Oral presentation, *XXII Congress of the International Primatological Society*, Edinburgh, UK, 03–08 Aug 08

PETRŽELKOVÁ, KLÁRA J.; LUCIA BOBÁKOVÁ; LIZA A. MOSCOVICE; MWANAHAMISI I. MAPUA; HEIDO HASEGAWA; MICHAEL A. HUFFMAN; JANA PETRÁŠOVÁ; TARANJIT KAUR (2008) Successful introduction of zoo chimpanzees onto Rubondo Island (Tanzania). Poster, *25th EAZA Conference*, Anwerp, Belgium, 16–20 Sep 08

PETRŽELKOVÁ, KLÁRA J.; JANA PETRÁŠOVÁ; M. UZLIKOVA; M. KOSTKA; MICHAEL A. HUFFMAN; MWANAHAMISI I. MAPUA; LUCIA BOBÁKOVÁ; VLADIMÍR MAZOCH; JATINDER SINGH TARANJIT KAUR; DAVID MODRY (2010) Gastrointestinal parasites of indigenous and introduced primate species of Rubondo Island National Park, Tanzania with emphasis on Blastocytis infections. Abstract, *Primate Research* 26: 181. *International Primatological Society 23rd Congress*, Kyoto, Japan, 12–18, Sep 10

MSINDAI, JOSEPHINE N. (2015) Nest site selection in the chimpanzees of Rubondo Island, Tanzania. Poster, *6th meeting, European Federation for Primatology & XXII API Congress*, Rome, Italy, 25–28 Aug 15. *Folia Primatologica* 86, 326

MSINDAI, JOSEPHINE N.; VOLKER SOMMER; CHRISTIAN ROOS (2015) The chimpanzees of Rubondo Island: Genetic data reveal their origin. Poster, *6th meeting, European Federation for Primatology & XXII API Congress*, Rome, Italy, 25–28 Aug 15. *Folia Primatologica* 86: 327

MSINDAI, JOSEPHINE N. (2021) The Rubondo Island population 50 years post release. Oral presentation, *Global Primatology Conference*, Central Washington University, USA, 25–26 Mar 21

MSINDAI, JOSEPHINE N. (2021) Rubondo Island chimpanzees. Invited online presentation, *Evolutionary Aquatic Ecology Seminar Series*, Bern University, 10 Mar 21

Unpublished Reports and Appraisals

(Additional grey literature likely exists. RINP = Rubondo Island National Park. Frequent recipients of reports: CAWN = College of African Wildlife Management, Tanzania; COSTECH = Tanzania Commission for Science and Technology; CSA = Climate-smart Agriculture Guideline, Tanzania; FZS = Frankfurt Zoological Society; TANAPA = Tanzania National Parks Authority; TAWIRI = Tanzania Wildlife Research Institute)

BORNER, MARKUS (1971) *Vegetation of Rubondo.* Report to FZS

EASTON, ENNETT R. (1974) *Preliminary Survey of the Ticks from Rubondo Island Forest Reserve, Geita District, Mwanza Region.* VIC, Mwanza

MATSCHKE, WOLFGANG (1974) *Annual Report 1973 of the Rubondo Island Game Reserve.* Game Division Mwanza

BORNER, MARKUS (1978) *Preliminary Survey of RINP.* Report to FZS

BORNER, MARKUS (1978) *Ecological Survey Progress Report.* Report to FZS

BORNER, MONICA (1980) *Ecological Survey, Rubondo Island National Park.* Report to FZS

MBANO, B.N.N.; H.J. MWAGENI (1987) *An Appraisal of the Wild Animal Populations in the RINP.* Report to CAWN

STRATON, LISA (1992) *RINP Bird Study Proposal*

Anonymous (1993) *The History and Current Status of the Black Rhinoceros in RINP, Tanzania.* Report to FZS

ERIKSMOEN, BRANDI (1993) *Elephant Ecology and Conservation on Rubondo Island.* Report to CSA

MÜLLER, GUIDO (1995) *Schimpansen auf Rubondo-Island, Tanzania: eine Pilotstudie.* Anthropologisches Institut, Universität Zürich

OMMANEY, DOUGLAS (1998) *Chimpanzee Status Survey and Habituation Project: Progress Report 1998.* Report to FZS

PUSEY, ANNE (1998) *Rubondo Chimp Project.* Report to FZS

ROBINSON, JOHANNA; PAUL ROBINSON (1998) *Chimpanzee Status Survey and Habituation Project: Progress Report 1998.* Report to FZS

HUFFMAN, MICHAEL A. (2000) *Evaluation of the RINP Chimpanzee Habituation Project.* Japan

MOSCOVICE, LIZA R.; MICHAEL A. HUFFMAN (2001) *Ecology of an Introduced Group of Chimpanzees at RINP, Tanzania.* Preliminary Report. University of Wisconsin-Madison

REITSMA, NICOLE (2001) *A Behavioural Study of Vervet Monkeys on Rubondo Island.* Report to CSA

SOCHA, ANDREA (2001) *Nesting Site Preferences of Chimpanzees On RINP.* Wildlife Ecology and Conservation. Report to CSA

KIWANGO, YUSTINA A. (2002) *A Survey of Ecological Issues Facing the Park.* Report to TANAPA

KIWANGO, YUSTINA A.; HOBOKELA R. MWAMJENGWA; STERIA R. NDAGA (2005) *Rubondo Island National Park: Draft Report on Ethnoecology Study.* Report to TANAPA

TANAPA (2003) *General Management Plan for Rubondo Island National Park 2003–2013.* Arusha, Tanzania

HUFFMAN, MICHAEL A.; KAUR, TARANJIT; SINGH, JATINDER; PETRŽELKOVÁ, KLÁRA; ISSA, MWANAHAMISI; RESTIS, EVA; BOBÁKOVÁ, LUCIA; SZEKELY, BRIAN; MAZOCH, VLADIMÍR (2006) *Rubondo Island and Mahale Mountains Chimpanzee Initiative: Integrating Socio-Ecological Research and Conservation Medicine.* Report to Costech

PETRŽELKOVÁ, KLÁRA J.; LUCIA BOBÁKOVÁ; VLADIMÍR MAZOCH (2010) *Rubondo Island and Mahale Mountains Chimpanzee Initiative: Integrating Socio-Ecological Research and Conservation Medicine.* Report to TAWIRI

TAWIRI (2018) *Tanzania Chimpanzee Conservation Action Plan 2018–2023.* Arusha, Tanzania

Student Theses (Undergraduate, Master, PhD)

MOSCOVICE, LIZA R. (2006) *Behavioral Ecology of Chimpanzees (Pan troglodytes) on Rubondo Island, Tanzania: Habitat, Diet Grouping and Ranging at a Release Site.* PhD thesis. University of Wisconsin-Madison

HECZKOVÁ, KATEŘINA (2007) *Parasite Infections of Vervet Monkeys in an Isolated Island Ecosystem.* Bachelor thesis. Masaryk University

REPTOVÁ, Z. (2008) *Morphological Variability of Eggs of Trichuris sp. in Primates.* Unpublished manuscript as part of a Bachelor thesis. University of Veterinary and Pharmaceutical Sciences Brno

REED-SMITH, JANICE (2010) *An Assessment of Seasonality and Shoreline Characteristics Associated with Latrine Site Use by Spotted-Necked Otters (Lutra maculicollis, Lichtenstein 1835) with a Preliminary Report on Spotted-Necked Otter Behavior on Rubondo Island National Park, Tanzania.* Master's thesis. George Mason University

AMULIKE, BRIDGET (2011) *Enhancing Tourist Opportunities to View Spotted-Necked Otters (Lutra maculicollis) at Rubondo Island National Park: Can the Apriori Location of Latrines Simplify Identifying the Best Viewing Areas?* Master's thesis. Frostburg State University

STEVENS, SADIE S. (2011) *Flagship Species, Tourism, and Support for Rubondo Island National Park, Tanzania.* PhD thesis. University of Massachusetts

GARA, JOHN IGNASE (2018) *The Effect of the Litter on the Lake Shoreline: A Case of Rubondo Island National Park, Tanzania.* Master's thesis. The Open University of Tanzania

MSINDAI, JOSEPHINE N. (2018) *Chimpanzees of Rubondo Island: Ecology and Sociality of a Reintroduced Population.* PhD thesis. University College London

WILLIAMS, CATRYN (2018) *Evolution of the Primate Gut Microbiome.* PhD thesis. University College London [pp. 125–159]

Popular Writing and Tourism Stories

Anonymous (1966) Chimps with a problem. *The Standard* (Tanzania) (17 Jun)

GRZIMEK, BERNHARD (1966) Operation chimpanzee: youngster would not leave us. *Sunday News* (Tanzania) (18 Dec)

GRZIMEK, BERNHARD (1967) Operation chimpanzee: giant chimp starts dancing. *Sunday News* (Tanzania) (15 Jan)

KADE, ULRICH (1967) Ich lebe mit den Schimpansen am Viktoriasee. *Das Tier* 6/7, 31

MATSCHKE, WOLFGANG (1976) Rubondo – Erinnerung an eine Naturschutzinsel. *Der Deutsche Forstmann* 2, 24–25

BORNER, MONICA (1980) Rubondo – ein Nationalpark mausert sich. *Das Tier* 10, 6–9

MWAGENI, H.J. (1991) Rubondo INP. *Kakakuona* 3, 3

REED-SMITH, JANICE; TUA S. KIHEDU; HOBOKELA MWAMJENGWA; MAJALIWA M. MUHABI (2010) Africa's otter surprise: Flagship for clean water. *SWARA Magazine* 1, 58–60

WATT, SUE (2014) Afrika's ark. *Wildlife Extra* (01 Aug) [www.suewatt.co.uk/docs/rubondo_feature_wt_aug_2104.pdf]

ROWAN, ANTHEA (2015) Tanzania's sanctuary for threatened animals. *BBC Travel* (15 Sep) [bbc.com/travel/story/20150915-tanzanias-very-own-noahs-ark]

RUNNETTE, CHARLES (2015) Africa's secret fantasy-island safari. *The Wall Street Journal* (20 Oct) [wsj.com/articles/africas-secret-fantasy-island-safari-14453 66941]

SCHÜRMANN, FELIX (2017) Grzimeks Afrika. …zwischen westlichem Naturschutzkonzept und kolonialen Klischees. *Zeitgeschichte-online* (13 Mar) [dev.zeitgeschichte-online.de/film/grzimeks-afrika]

STEWART, CATRINA (2017) Victoria's enchanted isle. *Nomad Magazine* (14 Sep) [nomad.africa/victorias-enchanted-island]

Asiliaafrica (2018) Chimpanzees. *BBC Wildlife Magazine* (26 Sep)

STEWART, JESSICA (2018) Descendants of rescued zoo chimpanzees thrive in Tanzanian jungle. *My Modern Met* (15 Mar) [mymodernmet.com/george-turner-rubondo-island-chimpanzee-pictures]

CALKIN, JESSAMY (2019) The real Jurassic Park: Inside story of the wild zoology experiment that could have gone horribly wrong. *The Telegraph* (19 Oct)

TISDALL, NIGEL (2019) Island of the apes: meeting the lucky chimps of Lake Victoria. *Financial Times* (04 Jan) [ft.com/content/886f9d36-0a9e-11e9-a242-6043097d0789]

DINTER, MARCO (2021) ZGF-Historie: Wie Grzimek die Schimpansen zurück nach Afrika brachte. *Gorilla (Magazin der Zoologischen Gesellschaft Frankfurt von 1858 e.V.)* 3, 24–31

MSINDAI, JOSEPHINE N. (2021) Ein Experiment mit gutem Ausgang: Die Schimpansen erobern Rubondo. *Gorilla (Magazin der Zoologischen Gesellschaft Frankfurt von 1858 e.V.)* 3, 32–36

RABEL, KASIA (2022) An island where chimpanzees rule. *FZS* (02 Nov) [https://fzs.org/en/news/an-island-where-chimpanzees-rule/]

Documentaries

GRZIMEK, BERNHARD (1966) *Ein Platz für Tiere.* TV episode, channel 1, ARD, West-Germany (47 min, 15 Nov)

Index

For Product Safety Concerns and Information please contact our EU
representative GPSR@taylorandfrancis.com
Taylor & Francis Verlag GmbH, Kaufingerstraße 24, 80331 München, Germany